AF765182

Advanced Technology in Neurosurgery

Edited by F. Pluchino and G. Broggi

With the Assistance of C. L. Solero and M. Fornari

With 152 Figures

Springer-Verlag
Berlin Heidelberg New York
London Paris Tokyo

From the Meeting on "Advanced Technology in Neurosurgery", held in Milan on May, 29 – June, 1, 1985

Prof. FRANCO PLUCHINO, M.D.
Prof. GIOVANNI BROGGI, M.D.
C. L. SOLERO, M.D.
M. FORNARI, M.D.
Department of Neurosurgery
Istituto Neurologico "C. Besta"
Via Celoria, 11
I-20133 Milan, Italy

ISBN-13:978-3-642-83125-6 e-ISBN-13:978-3-642-83123-2
DOI: 10.1007/978-3-642-83123-2

This work is subject to copyright. All rights are reserved, whether the whole or part of the material is concerned, specifically those of translation, reprinting, re-use of illustrations, recitation, broadcasting, reproduction on microfilms or in other ways, and storage in data banks. Duplication of this publication or parts thereof is only permitted under the provision of the German Copyright Law of September 9, 1965, in its version of June 24, 1985, and a copyright fee must always be paid. Violations fall under the prosecution act of the German Copyright Law.

© Springer-Verlag Berlin Heidelberg 1988
Softcover reprint of the hardcover 1st edition 1988

The use of registered names, trademarks, etc. in this publication does not imply, even in the absence of a specific statement, that such names are exempt from the relevant protective laws and regulations and therefore free for general use.

Product Liability: The publishers can give no guarantee for information about drug dosage and application thereof contained in this book. In every individual case the respective user must check its accuracy by consulting other pharmaceutical literature.

2122/3130-543210

List of Contributors

Anderson, R.E., Department of Neurologic Surgery,
Mayo Medical School, Mayo Clinic, Rochester, MN 55905, USA

Askienazy, S., Service de Médecine Nucleaire, Hôpital St. Anne,
1 rue Cabanis, F-75014 Paris, France

Avanzo, R.C., Department of Medical Physics, City Hospital,
viale Rodolfi, I-36100 Vicenza, Italy

Benedetti, A., Department of Neurosurgery, City Hospital,
viale Rodolfi, I-36100 Vicenza, Italy

Betti, O.O., Servicio de Neurocirurgia Tridimensional y Radiocirurgia
de Institutos Médicos Antartida,
Rivadavia 4980 (1424), Buenos Aires, Argentina

Berenstein, A., Department of Neuroradiology,
New York University Medical Center,
550 First Ave., New York, NY 10016, USA

Brant-Zawadzki, M., Department of Neuroradiology, Moffit Hospital,
University of California at San Francisco,
San Francisco, CA 94143, USA

Brock, M., Department of Neurosurgery, Free University of West Berlin,
Steglitz Medical Center,
Hindenburgdamm 30, D-1000 Berlin 45, FRG

Broggi, G., Department of Neurosurgery, Istituto Neurologico "C. Besta",
Via Celoria, 11, I-20133 Milan, Italy

Cerchiari, U., Department of Physics, Istituto Nazionale dei Tumori,
Via Venezian, 1, I-20133 Milan, Italy

Chierego, G., Department of Medical Physics, City Hospital,
viale Rodolfi, I-36100 Vicenza, Italy

Chodkiewiecz, J.P., Service de Neurochirurgie, Hôpital St. Anne,
1 rue Cabanis, F-75014 Paris, France

Choi, I.S., Department of Neuroradiology,
New York University Medical Center,
550 First Ave., New York, NY 10016, USA

Clodic, R., Hôpital St. Anne, Service de Neurochirurgie,
1 rue Cabanis, F-75014 Paris, France

Colombo, F., Department of Neurosurgery, City Hospital, viale Rodolfi, I-36100 Vicenza, Italy

Cosman, E.R., Department of Physics, Massachusetts Institute of Technology, Cambridge, MA 02139, USA

Derechinsky, V.E., Servicio de Neurocirurgia Tridimensional y Radiocirurgia de Institutos Médicos Antartida, Rivadavia 4980 (1424), Buenos Aires, Argentina

Epstein, F., Department of Neurosurgery, New York University Medical Center, Division of Pediatric Neurosurgery, 550 First Ave., New York, NY 10016, USA

Ernst, H., Department of Radiation Therapy, Free University of West Berlin, Steglitz Medical Center, Hindenburgdamm 30, D-1000 Berlin 45, FRG

Flamm, E., Department of Neurosurgery, New York University Medical Center, 550 First Ave., New York, NY 10016, USA

Franzini, A., Division of Neurosurgery, Istituto Neurologico "C. Besta", Via Celoria, 11, I-20133 Milan, Italy

Fraysse, B., Department of Otorhinolaryngology, Hôpital PURPAN, Place du Docteur Baylac, F-31059 Toulouse Cedex, France

Galmarini, D.H., Departamento de Radiocirurgia del Centro de Radioterapia del Hospital Espanol, Buenos Aires, Argentina

Giorgi, C., Division of Neurosurgery, Istituto Neurologico "C. Besta", Via Celoria, 11, I-20133 Milan, Italy

Hosobuchi, Y., Department of Neurosurgery, School of Medicine, University of California at San Francisco, San Francisco, CA 94143, USA

John, E.R., Brain Research Laboratories, Department of Psychiatry, New York University Medical Center, 550 First Ave., New York, NY 10016, USA

Jori, G., Division of Neurosurgery, General Hospital, Department of Biology, University of Padua, I-Padua, Italy

Kelly, P.J., Department of Neurosurgery, Mayo Medical School, Mayo Graduate School of Medicine, Mayo Clinic, Rochester, MN 55905, USA

Kelly, W.M., Department of Neuroradiology, Moffit Hospital, University of California at San Francisco, San Francisco, CA 94143, USA

Laws, E. R. Jr., Department of Neurologic Surgery, Mayo Medical School, Mayo Clinic, Rochester, MN 55905, USA

Lazorthes, Y., Service de Neurochirurgie, Clinique Neuro-chirurgicale, C.H.U. Rangueil,
Chemin du Vallon, F-31054 Toulouse Cedex, France

Malis, L. I., Department of Neurosurgery, Mount Sinai School of Medicine, 1176, 5th Ave. – Annenberg Building 25–86, New York, NY 10029, USA

Marchetti, C., Department of Medical Physics, City Hospital, viale Rodolfi, I-36100 Vicenza, Italy

Mingrino, S., Division of Neurosurgery, General Hospital, Department of Biology, University of Padua, I-Padua, Italy

Munari, C., INSERM U97, 2 ter rue d'Alesia, F-75014 Paris, France

Musolino, A., Service de Neurochirurgie, Hôpital St. Anne, 1 rue Cabanis, F-75014 Paris, France

Newton, T. H., Department of Neuroradiology, Moffit Hospital, University of California at San Francisco, San Francisco, CA 94143, USA

Norman, D., Department of Neuroradiology, Moffit Hospital, University of California at San Francisco, San Francisco, CA 94143, USA

Oppel, F., Department of Neurosurgery, Free University of West Berlin, Steglitz Medical Center, Hindenburgdamm 30, D-1000 Berlin 45, FRG

Ostertag, Chr. B., Abteilung für Stereotaktische Neurochirurgie, Neurochirurgische Klinik der Universität des Saarlandes, D-6650 Homburg/Saar, FRG

Pannek, H. W., Department of Neurosurgery, Free University of West Berlin, Steglitz Medical Center, Hindenburgdamm 30, D-1000 Berlin 45, FRG

Penn, R. D., Department of Neurosurgery, Rush-Presbyterian-St. Luke's Medical Center, 1753 W. Congress Parkway, Chicago, IL 60612, USA

Pozza, F., Department of Radiotherapy, City Hospital, viale Rodolfi, I-36100 Vicenza, Italy

Prichep, L. S., Brain Research Laboratories, Department of Psychiatry, New York University Medical Center, 550 First Ave., New York, NY 10016, USA

Ransohoff, J., Department of Neurosurgery,
New York University Medical Center,
550 First Ave., New York, NY 10016, USA

Rea, G., Neurosurgical Department, University Hospital,
CH-8091 Zürich, Switzerland

Siegfried, J., Neurosurgical Department, University of Zürich,
Present address: Klinik im Park, Seestr. 220,
CH-8002 Zürich, Switzerland

Soulier, M. J., Department of Otorhinolaryngology, Hôpital PURPAN,
Place du Docteur Baylac, F-31059 Toulouse Cedex, France

Stein, B. M., Department of Neurological Surgery, Columbia University,
College of Physicians and Surgeons, and Presbyterian Hospital,
New York, NY 10032, USA

Verdie, J.-C., Service de Neurochirurgie, Clinique Neuro-chirurgicale,
C. H. U. Rangueil,
Chemin du Vallon, F-31054 Toulouse Cedex, France

Vincent, M., Department of Otorhinolaryngology, Hopital PURPAN,
Place du Docteur Baylac, F-31059 Toulouse Cedex, France

Wharen, R. E. Jr., Department of Neurologic Surgery, Mayo Medical School,
Mayo Clinic, Rochester, MN 55905, USA

Zampieri, P., Division of Neurosurgery, General Hospital,
Department of Biology, University of Padua,
I-Padua, Italy

Zanardo, A., Institute of Applied Mechanics, University of Padova,
I-Padova, Italy

Preface

Neurosurgery has made enormous progress in the past 15 years, in a completely different way to other fields of surgery. What has characterized the originality of the progress in neurosurgery is the comprehensive modification of the operative technique itself, both regarding the equipment at our disposal and the strategy behind the planning of the surgical operation. These modifications have been so radical and substantial that experience accumulated, even in the recent past, by distinguished doctors who have preceded us in this work is often rendered of little use. Progress obtained in other disciplines, on the other hand, is generally only the result of continuity in surgical experience.

If functional neurosurgery has opened new horizons and prospects, hard to foresee, microsurgical techniques have played a decisive role in promoting this development. Their rapid spread came about first because of the right rate of good clinical results reported by their forerunners, and then by the steady confirmation of their value in numerous operations carried out by many different surgeons.

The indications for surgery in space-occupying lesions and vascular malformations of the brain and spinal cord have now been broadened to include locations once considered surgically inaccessible. The clinical results obtained nowadays are of such a high standard that past results must be considered unacceptable.

Microsurgery has not only given us more adequate means of performing neurosurgical operations while preserving important nervous structures, but it has also given us a thorough knowledge of surgical anatomy and the morphology of lesions in the central nervous system. Thus the operating microscope, coupled with the bipolar coagulator and the different microsurgical instruments, must be considered the most significant technological improvement of the 1970s in neurosurgery. The technological and scientific progress of neuroanesthesia and neuroreanimation have also been of great help to this progress. In recent years technology has given new tools to the neurosurgeon, offering new prospects, but at the same time demanding experimental and clinical demonstration of the usefulness of these instruments. Lasers, the ultrasound aspirator, appliances for monitoring evoked potentials, Doppler flowmeter, echo tomography, and stereotactic instruments compatible with computed tomographic scan and nuclear magnetic resonance are only some of the new devices available. Our task is to find out whether all these instruments, astonishing in their mechanical conception, really render appropriate, concrete assistance in our work, and whether they are therefore essential.

We must not ignore progress, but everything new must be evaluated without preconceptions and with caution before it is given the name of "progress." This is even more important because of the high cost of new equipment; the clinical results should justify the financial obligation for the purchase of the instruments. The increased cost to the health service, a burden to all taxpayers, must at least be counterbalanced by an effective improvement in the diagnostic and therapeutic services supplied to the patient.

Milan, Italy, 1987

Franco Pluchino
Giovanni Broggi

Contents

New Trends in Microsurgery and Applied Technology

L. I. Malis

I shall begin with a bit of the history before going on to what is new. Actually, Professor Yasargil and I did a bit of discussing since we speak together on the same topic and decided on how to divide it up. Since I am older than he is, I took the history.

The first applications of the actual binocular microscope to human surgery began almost 70 years ago with the first publications by Holmgren who did repair of otosclerosis with a true binocular microscope (Fig. 1). Simple monocular devices had been used before that, but Holmgren's work published in 1922 was the first true microsurgery. The otologists developed microsurgical techniques extensively over the years following Holmgren's innovation. By 1953 Littman of the Zeiss Company had developed the Zeiss Opton (Fig. 2) still sold as the OPMI-1, which has been the standard surgical scope for a generation. This microscope also was widely used in the physiological laboratories. My own experience goes back to working in the research laboratory with the microscope for many years doing animal work, knowing that of course it could not be used in neurosurgery. I had watched the otologists work and I was certain that contamination was continuous and infection inevitable. So much for one's planning ahead.

Finally, 20 years ago (in May 1965) I was confronted by what I found to be an unremovable craniopharyngioma which I had exposed through a generous craniotomy. In the depth of the exposure I had only a little slit between the optic nerve and the carotid through which the lesion could be approached (Fig. 3). The tumor was hard,

Fig. 1. Holmgren's original microscope, 1922

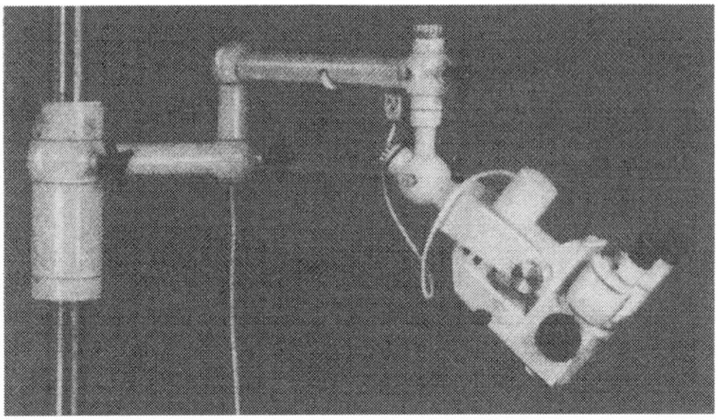

Fig. 2. The Zeiss OPMI-1 as first introduced

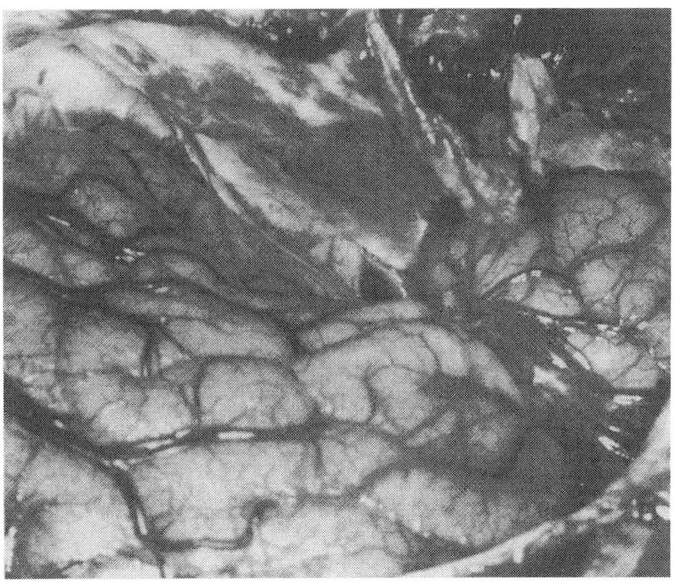

Fig. 3. First microsurgical removal of craniopharyngioma, 1965. Right fronto-pterional approach with opening of Sylvian fissure. The tumor was removed through the triangular space between the optic nerve and carotid artery

not cystic, displacing the chiasm anteriorly and upward, lifting the 3rd ventricle and pushing down even into the posterior fossa. I brought in the microscope from the otologic room next door and had my instruments brought up from the animal laboratory and some time late that night finished the removal. This young man is still alive without a recurrence. From that time on, 20 years ago, I started to become a microneurosurgeon.

In the mid sixties Kurze, on the west coast of the United States, had begun to use the microscope for posterior fossa neurosurgery. On the east coast, Jacobson had brought his pioneering microvascular technology to Donaghy's laboratory. By 1966,

Fig. 4. Microneurosurgical course at Mount Sinai, 1969

Fig. 5. Microneurosurgical course at Mount Sinai, 1965. Dr. Donaghy on the left, Dr. Yasargil on the right

Donaghy's department had become the training and research center, and Yasargil got his microsurgical laboratory training there. During his year with Donaghy, Yasargil took time to visit me and complete my conversion to microneurosurgery.

By 1968, we had gotten to the point where we were giving practical courses in microneurosurgery, first in Zürich. In my department at Mount Sinai by 1969, 40 microscopes were gathered, each supplied with anesthetized rabbits, and 160 neurosurgeons taking the practical course, with Donaghy and Yasargil participating with me in the teaching program (Figs. 4, 5). After so many years of delay, the last decade has seen the explosive microrevolution, with the microtechnique now part of virtually every neurosurgical center.

The old OPMI stand had been modified by adding pieces over the years to somewhat improve its mobility and adaptability. It is finally being replaced by 2 very dif-

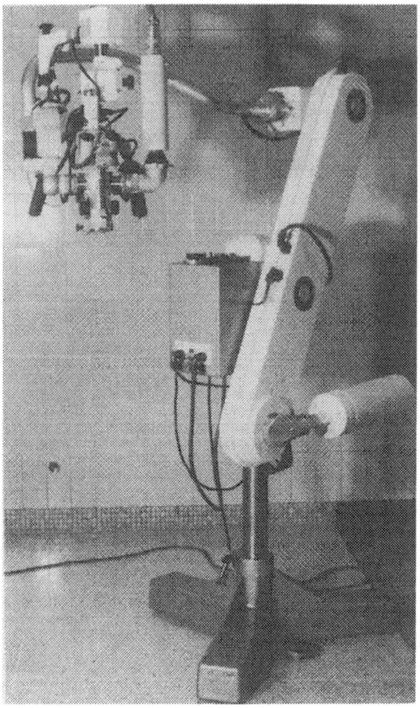

Fig. 6. Yasargil's Contraves prototype

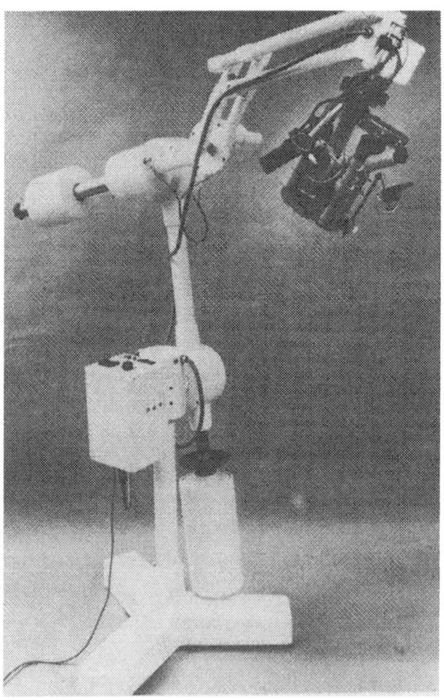

Fig. 7. Commercial Contraves microscope stand

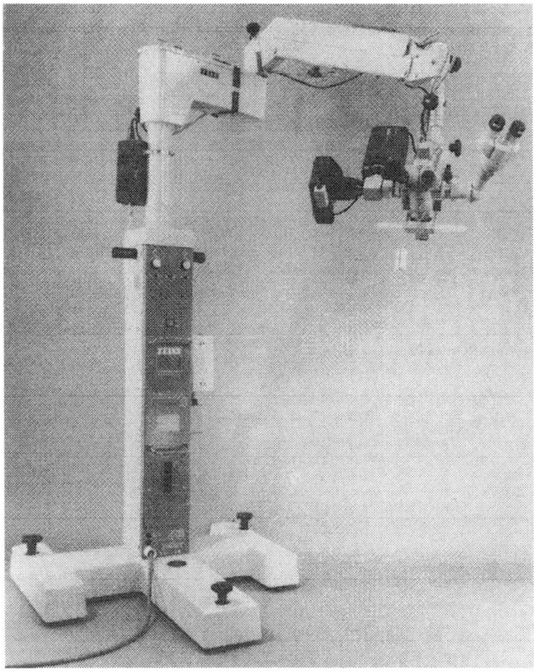

Fig. 8. The Zeiss Universal stand with dual fiberoptic lighting, OPMI-1 head, 35 mm camera, Panasonic color camera and binocular assistant's viewer

ferent stands. The Contraves prototype stand (Fig. 6) which Yasargil uses and about which he will speak further, is not commercially available. A much heavier higher inertia stand, the present commercial Contraves model, was marketed by Zeiss (Fig. 7) which I dislike thoroughly. The new Zeiss Universal stand I think is very useful. It is a microscope relatively easy to handle and it moves around with reasonable ease. It allows you to go from position to position without locking or unlocking holding its tension quite decently with minimal pressure to move it. So it is a very simple stand which will cover almost all positions. What is involved is really 2 different types of microtechnique. I am a sitting down surgeon who rarely moves the scope and then only a few degrees, and if I move a larger amount, I move everything slowly and then stay there a long time. Yasargil is a very rapid microsurgeon who does everything from the standing position and moves around tremendously and rapidly. His technique requires a scope like the Contraves, free floating, able to move to all parts. In contrast, I rarely move a scope except in one plane. These are 2 diametrically different techniques which fortunately both give good results.

The result is that I use my old OMPI-1 microscope head, with a pair of Vertolux fiberoptic light sources, now on the Universal stand. The assistant scope as well as the 35 mm camera and the television camera are mounted on the beam splitter (Fig. 8). No alteration is needed as one goes from one type of case to another. With my style of microsurgery I am now satisfied that this is the best available unit.

Now the whole design of an operating room around the microscope is exemplified by the new Siemens Company system which floats the microscope from a giant ceiling mount with everything else. The difficulty with this is that you are

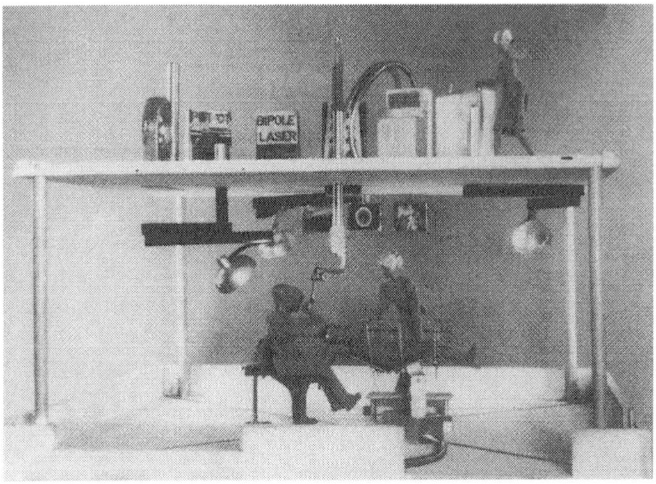

Fig. 9. Hypothetical two-story operating room for future development

pinioned into the chair with your shoes off to operate the pedals like an organist with everything running through buttons and you cannot move anything freely. While some are quite happy with this, to me this is a prison and an obstruction. Additionally, if one motor or one switch fails, you have to close the operating room, not just move in another microsope.

In general, I much prefer the mobility, freedom and stability of free standing floor microscopes to ceiling mounts. I suggested, in jest, because of problems with vibration with ceiling microscopes, that we should build a two-story operating room and put the microscope on the floor of the second story and let it hang downward through a hole. I made up this little model in which we have got everything upstairs (Fig. 9). It has never been built and almost certainly never will be. But it is a concept of one of the ways to go in the future. I show it to the architects every now and then to upset them when they plan new buildings.

With increasing complexity in the types of procedures, the microscopes and the ancillary equipment, one should mention the absolute requirement of a highly knowledgeable and qualified neurosurgical operating room nurse. It of course is much easier if the nurse has 3 arms instead of the usual 2 (Fig. 10).

In the era before microsurgery, many of the operating rooms were built with windows which decreases your ability to look into the microscope because you have too bright a light in the room. Additionally, windows produce distractions and the elimination of the window from the operating room was one of the important things. In our old building, before we built without windows, I wound up painting all the windows black.

The modern operating room for me requires preparation rooms where the patient is placed on the table for prepping, where all the equipment is neatly organized and available outside of the operating room so that none of this preparation process occurs inside the OR endangering sterility. From the preparation room the patient is wheeled completely ready for surgery into the operating room, and then the team

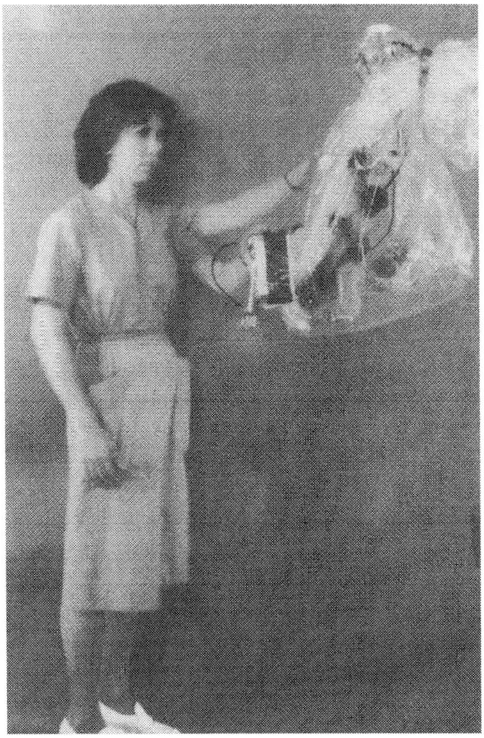

Fig. 10. Hypothetical three-armed nurse for future development

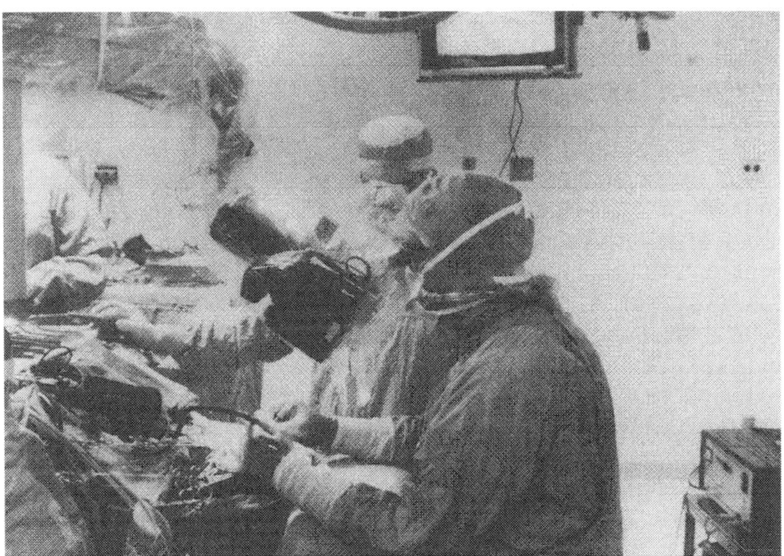

Fig. 11. Operating with a patient in the supine position

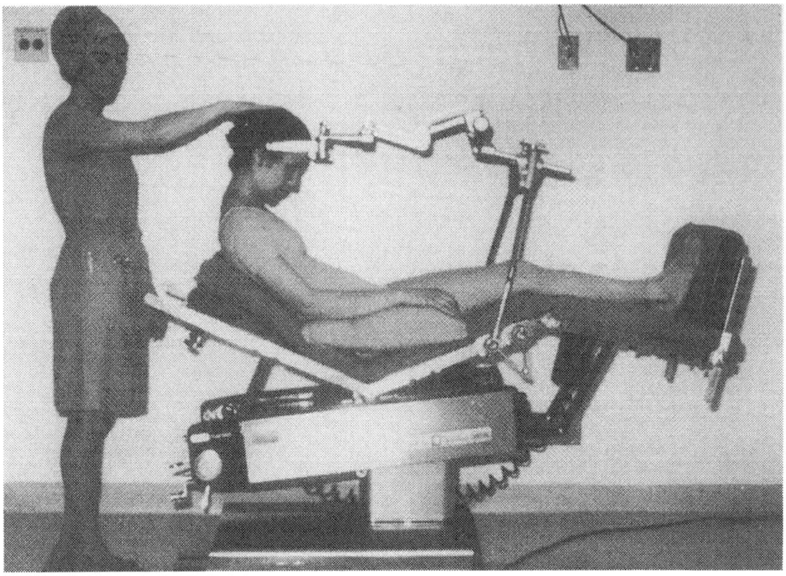

Fig. 12. The low-sitting position. Note minimal elevation of back of table

can go to work. We begin with four-diameter loupes and fiberoptic headlamps until the dura mater is exposed and we then go on to the microscope.

I usually do everything with the patient in one of 3 positions; first, with the patient lying supine in the standard back position (Fig. 11), second, in the semisitting position with the table only slightly elevated so that the patient is down at a low level (Fig. 12) so I can sit comfortably behind the patient without pushing the microscope way up in the air. This position requires the doppler monitor to check for the possibility of air entering the patient's venous system while end expiratory CO2 monitoring indicates the degree of seriousness of embolization. The other position I use is the prone-oblique-position at 45° which allows you to expose the areas anywhere from the base of the skull or the posterior fossa all the way down to the sacrum or anywhere in between (Fig. 13). It allows you to sit comfortably behind the patient and to work at a 45° angle moving freely around with arm support and without bending over the patient, and yet providing free abdominal and respiratory movement so that the spine does not move in and out of the focal plane of the microscope permitting everything to stay in focus while you operate, even at high magnifications.

You will hear much more about the evoked potential equipment and the additional values it provides while you are working in the field. I have been using it rather continuously for some time. We have, for example, monitored more than 200 acoustic neuromas during surgery. However, our results statistically are identical with those we did not monitor, so that I am not sure of the ultimate value.

We have the television communication so that the neuroradiologist can show us a specific area from a film on his black and white camera. The radiologist uses the color picture coming down from the operating room so that we can compare our anatomical findings. One of the major things that has been different in microsurgery

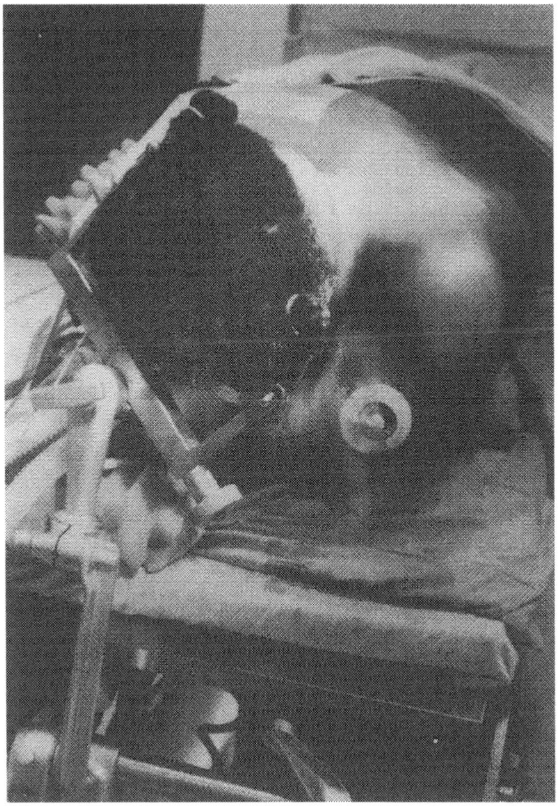

Fig. 13. Prone oblique position

is the anatomical knowledge that we have gained regarding the arachnoid planes and cisterns, and the fine vessel arrangements as well as the characteristic displacements that occur and our ability to predict and know them rather than to explore blindly.

Dr. Yasargil suggested to me that it might be good if I spent more time talking about coagulation than anything else, and since I am deeply interested in that and since it does not come up elsewhere in the program, I thought I would finish with a rather more detailed discussion of coagulation. An electrosurgical device with a spark generator was first brought into an operating room in 1910. It had a motor-driven rotating-disc as a static generator with a spark gap, a Leyden jar as a capacitor and a resonating coil. It put out a powerful discharge with which one could almost cook living tissue but which was not generally used. In 1928, W. T. Bovie, a physicist and one of the earliest biophysicists, developed an electrosurgical unit in which a spark gap generator provided an output of damped irregular waveforms of about a megahertz in frequency which permitted coagulation, and with a synchronous resonating circuit, also providing cutting ability. Cushing used this machine to revolutionize some of his removals in areas where bleeding had not been controllable before. Cushing's first publication began with a note by Bovie which is remarkably erudite now more than 65 years later.

Let me take a moment for basic principles. Direct current has a voltage level which stays constant through time. Alternating current on the other hand changes its direction with each side of the circuit changing from positive to negative, and the time it takes to do this is the frequency of the alternating current. Our ordinary power line current is either 50 or 60 cycles per second (Hertz). The amplitude up from the zero line is the voltage and this is the same as pressure in a hydraulic system. The current, which is measured in amperes, is a measure of how many electrons have passed through in a unit of time. The total power is the volts times the amperes but that applies only to direct current. The power line alternating current describes a sinusoidal wave form.

The amount of current that can flow at a given voltage depends upon the resistance of the material through which it flows. Resistance is a direct current term. Impedance is its equivalent for alternating current since other factors determine how alternating current can go through a material. Those are frequency and phase shift which I will not go into today because of time constraints. Constant current stimulators are the ones most frequently used for electrophysiology today. This is the diametric opposite of power distribution for our cities. If constant current were used, the current, whether one lamp or a million were turned on, would be the same. There would be no light from the million bulbs and the one bulb would disappear in a flash of smoke. With a constant voltage system, which is what is used, no matter how many bulbs you put in, no matter how many air conditioners run, since the voltage remains constant, all bulbs stay at the same brightness and appliances at the same level. For the coagulator nothing could be more important than constant voltage. In a constant current system, if you had saline over the field virtually nothing would be left for the coagulation at the tip since most of the current would flow through the saline, whereas in a constant voltage system, the closer the voltage is regulated, the lower will be the effect of the shunting of the saline and the more will the voltage and the power at the tip remain the same to go through the tissue.

The constant voltage is determined by the generator impedance. The lowest possible impedance in the generator gives you the best regulation of the voltage and the most constant voltage so that the current flowing is going to rise if the tissue impedance is low, and the voltage across the tissue will remain the same regardless of the impedance, regardless of the forceps area. Engineers will tell you that for transferring power, the generator impedance and the load impedance should match and that is perfectly true if your problem is the transfer of power with the highest efficiency. We are totally uninterested in efficiency. We have power to waste and what we want is for the power to be always available at the forcep tip. So the matching load idea in the system is for us totally without reason. Current must flow across at the tips of the forceps despite the fact that it is immersed in saline all the way, which requires a very low generator impedance as close as possible to a constant voltage system.

As Bovie pointed out, there is a difference between cutting and coagulating current. Cutting is performed with a constant sine wave, from a perfectly smooth resonant system. It also requires only a small tissue area to be involved. It is generally done with a fairly high voltage so that a tiny spark ahead of the electrode actually divides the tissue by molecular resonance. At one megahertz the macromolecules of tissue resonate smoothly in synchrony, fall apart and separate, and so the current cuts. Coagulating is done with bursts of damped waves, and a somewhat larger tissue area for the amount of power.

The important factor is the arrhythmicity. A coagulating waveform is an arrhythmic wave form where not 2 cycles are alike, where there can be no true resonance i.e. no molecular resonance and the lowest division of tissue. The two requirements for coagulating are an arrhythmic waveform and damping, which means the voltage will come in bursts and each burst will go downward and then recur again so that any resonance that takes place will be stopped between the bursts.

If you use a human and put an electrode on the brain and a ground plate on the back, the impedance between the two at about a megahertz is about 1500 ohms. The impedance between the tips of the bipolar forceps on the brain is about 80 ohms. Now under those circumstances, if you look at the power required, you realize that only 5% or so of the generator power is needed to coagulate the surface with the bipolar at that point on the brain than would be needed with the monopolar and ground plate.

When Greenwood first developed bipolar coagulation back in about 1939, he used it by putting one side of the forceps to each side of the Bovie machine and just disconnecting his ground plate so that he had a very high leakage from one side of his forceps to any ground on the patient. But if you look at it this way, the impedance between his forceps tips was 80 ohms, his impedance to ground at the best 1500 or so. Only 5% of the current was leaking away, i.e. 1/20th. So despite the fact that his system was totally without isolation, he only had a 5% leakage current and that isn't too bad and it did indeed work for him.

With the separate isolated bipolar unit, there is only the flow between the forcep tips and virtually none any other way, which does indeed prevent leakage current. If you use a unipolar forceps and just touch it with the Bovie, the best return path to ground is through the blood in the parent vessel and you may very well damage the parent vessel. With the bipolar, there is no flow into the parent vessel and, of course, you can seal just the branch you want, preserving the main trunk.

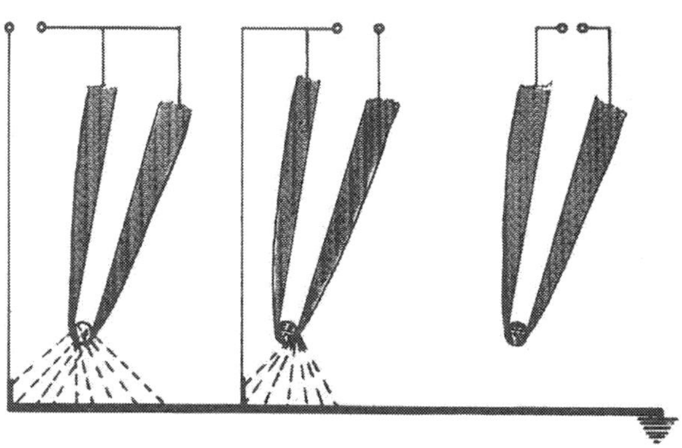

Fig. 14. On the *left* the unipolar coagulator. Current flows from the forceps to the dispersive electrode or ground plate. *Center section;* unisolated bipolar connection. Most of the current flows between the tips of the forceps but leakage current flows from the active electrode to the dispersive ground plate. On the *right* the isolated bipolar connection with the current flowing only between the forceps tips

To summarize this, you have the Bovie unit with the unipolar unit going to ground, the bipolar unisolated unit as Greenwood used it with 5% leakage in the human, and finally, the true isolated bipolar unit with all the power going through the forceps tips and no leakage to going ground (Fig. 14). Now when you go to repeat this experiment in the rat, you get a very different most misleading result. There have been papers in the literature saying that based on rat studies the leakage of some of the commercial units is too great to be safe. When you use a Bovie unit on a rat, the impedance is 150 ohms, so half the current will go to the ground plate; if you use an unisolated generator with bipolar forceps instead of 1/20 of the current as in the human. So you must take the rat leakage current studies and divide them by 10 in order to have any meaning for the human. To put it simply, the leakage current in any commercial bipolar, no matter how badly built, is not high enough to be a problem, no matter which company has made it. Reducing leakage current is helpful in avoiding interference with monitoring equipment and television cameras.

The way to determine leakage and its damage – if you are really worried about it – is to put one forcep blade of the bipolar on the surface of the rat brain and turn the power all the way up keeping it on for a minute or so. Then inject the live animal with methylene blue and if it doesn't stain, there is no significant damage. It is not the measurement of current that is important, it is what happens to the brain.

Another comparison between bipolar and unipolar coagulation: one side of the brain is stroked with the unipolar at the lowest power setting that will coagulate all the surface vessels; the other side with the bipolar under the same circumstances and then the methylene blue injection is given. There is indeed damage going deep with the unipolar which is the result of heat spread, and there is superficial damage from the bipolar, essentially the result of destroying the cortical vascular supply, and the

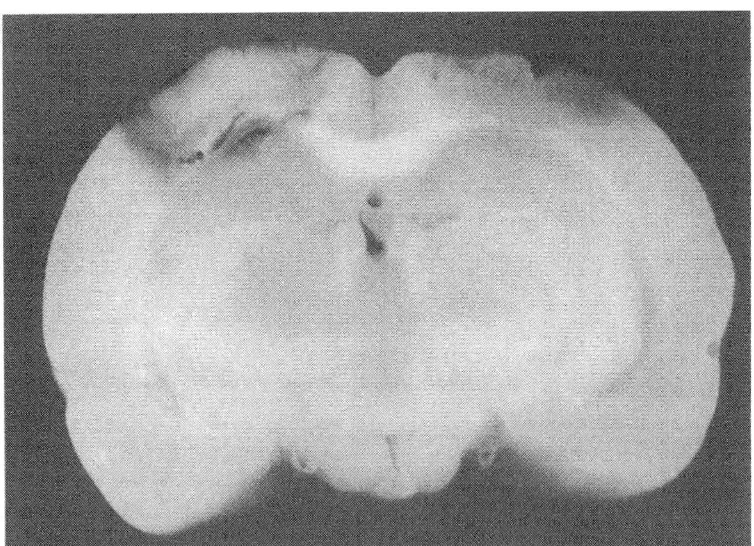

Fig. 15. On the *left* the deep necrosis caused by unipolar coagulation of surface vessels. On the *right* superficial change after bipolar coagulation of surface vessels

marked difference between the two is seen consistently (Fig. 15). Now if you set both unipolar and bipolar coagulators at the same power level you can blow the brain apart on the bipolar treated side because so much less power is required for coagulation than with the unipolar connection because the return, i.e., the ground plate, is at a distance. So the coagulators must be matched to the lowest power for each to coagulate for this to be meaningful.

The fact that we irrigate and do our coagulating under saline, since there is no current spread, allows us to keep tissue cool. I have had my assistant irrigate with a simple bulb syringe of saline. In the laboratory many years ago, when I had no assistant, I used to keep a needle cemented to my bipolar forceps and run saline through it (Fig. 16). We are going back to that again as progressive cost cutting leaves us without an assistant much of the time so that we now have available a forceps with an irrigating tube controlled to go on and off with the bipolar which makes the assistant unnecessary.

Different forceps sizes are essential, as part of your surgical armamentarium. You do not attempt to coagulate a vascular malformation with a fine pointed tip or you can indeed perforate it by getting too high a current density at the tip. You use a rounded, flat, broader blade which will allow you to shrink and bring down a vessel. The fine tip is superb for sealing a tiny branch of an artery which must be pre-

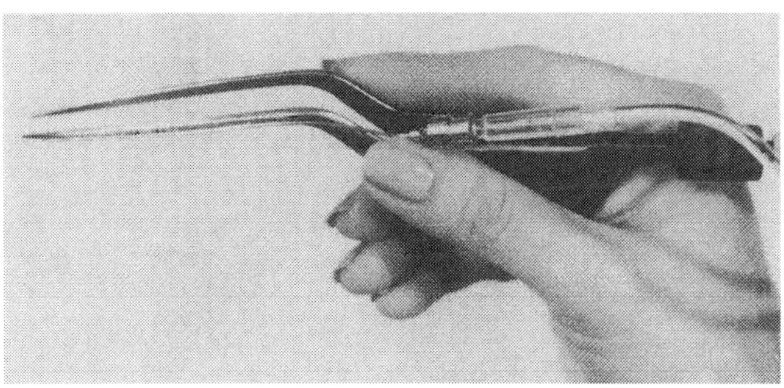

Fig. 16. Spinal needles cemented to side of bipolar forceps to provide irrigation

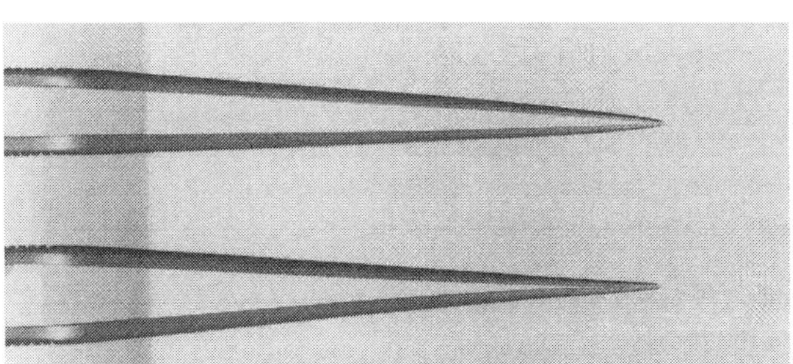

Fig. 17. Correct alignment of forceps blades in *upper half* of photograph, incorrect alignment with short circuiting proximal to the tips in the *lower* illustration

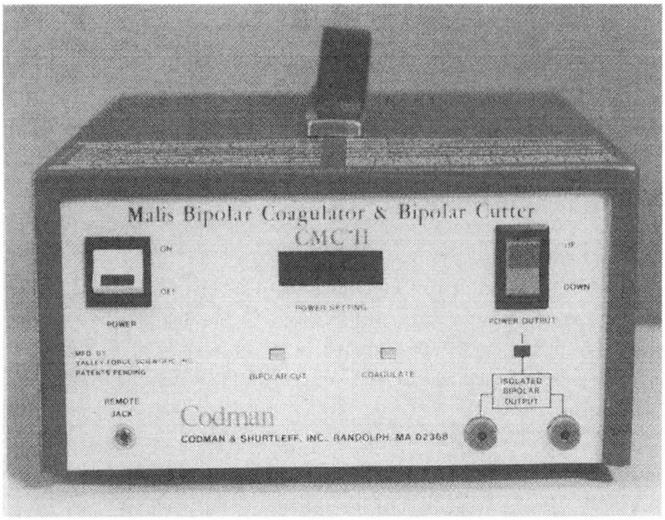

Fig. 18. The new solid state microprocessor controlled coagulator

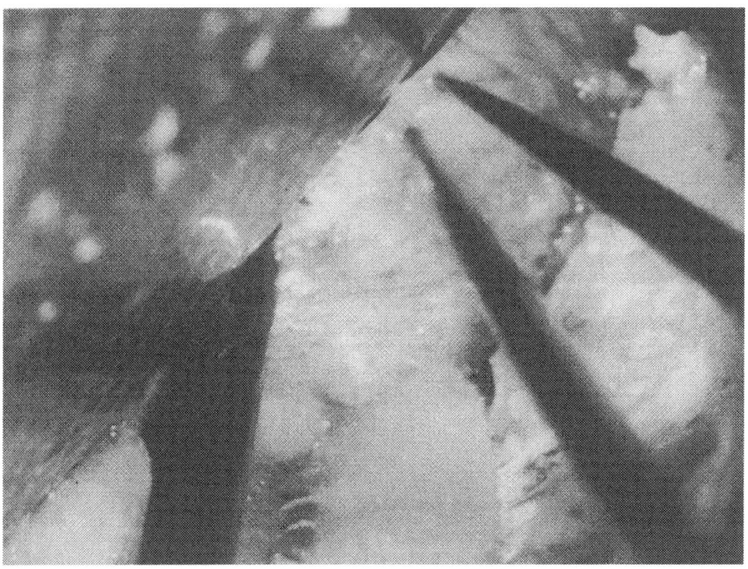

Fig. 19. A right acoustic neuromas with the forceps tips applied to the capsule

served, for working against a tumor capsule or taking a minute vessel off the facial nerve without damaging the nerve.

It is essential that the forceps not touch each other as you work because if the forceps tips are in contact, the current just goes through the metal and coagulates nothing. Forceps may come through from the factories with the blades straight. When

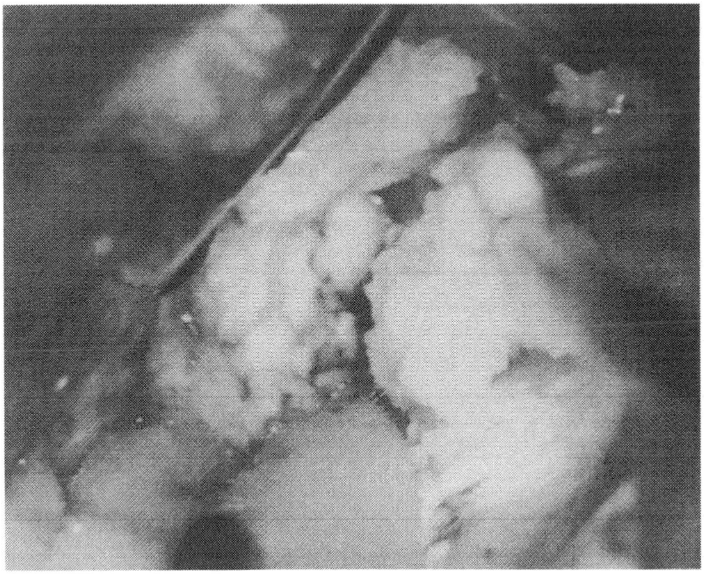

Fig. 20. Same patient as Figure 19. The bipolar cutting current is being used to core the tumor

they pick up a vessel at the tip, they are shorted well back of the tip. The forceps blades should be concave enough to meet so that you can hold the vessel without shorting (Fig. 17). It is also essential that the surface be clean and polished. I polish them myself with an electrical hone between cases since I do not trust them any other way.

The spark generator puts out a completely asynchronous highly random damped series of spikes in each burst. In some commercial solid state coagulators, a damped sinewave is generated. Because it does resonate in part, it has higher cutting and higher bursting of vessels. It is rhythmic and resonant despite the fact that it is damped. We were finally able with the advent of microprocessor technology to program a microprocessor to change the height and width of each spike of the damped wave, to change the interspike interval and so the frequency of each spike separately thus producing an asyncronous, arrhythmic damped wave with solid-state equipment. This has an advantage in that it does not have the high first spike of the spark generator. It begins with a gentle start and so there is almost no sparking when you use it on the tissue. This is the basis of my new solid-state unit which I personally find has a tremendous advantage over the old spark gap device. This is purely the result of microprocessor technology (Fig. 18).

Additionally, the new solid state bipolar unit incorporates a pure sinusoidal wave form for use as a bipolar cutting instrument. This again is a low impedence, relatively low voltage technique. It cuts by molecular resonance without the high voltage arc to the tissue. As such it works only under saline irrigation. It is effective and efficient for coring tumors but not so good for dividing avascular fibrous tissue. I now use the cutting mode in virtually all neoplasm surgery (Figs. 19, 20, 21). It does indeed decrease for me the difficulty and duration of the surgical procedures.

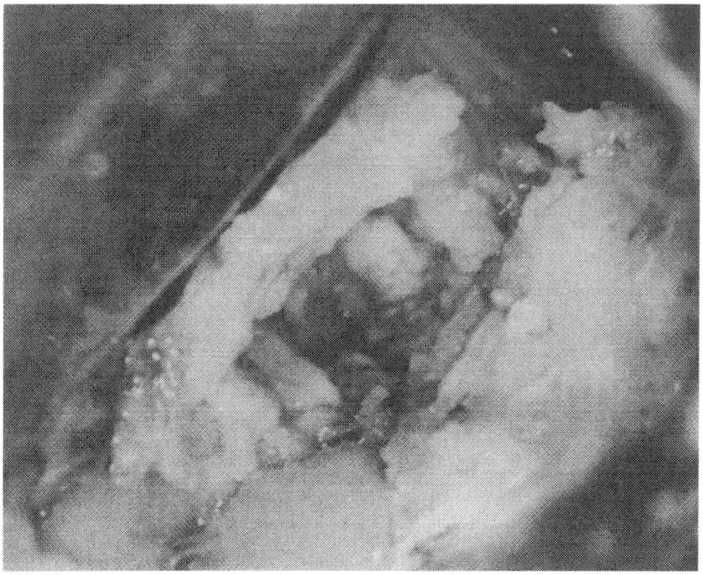

Fig. 21. Same patient as in Figs. 19 and 20. A large segment of the tumor has been removed bloodlessy with the bipolar cutting current

Finally, I have mentioned several times the problem with engineers and surgical instruments. Bovie brought to Cushing one day a pistol-grip with a trigger and a Bovie ball on the end of it and suggested that Cushing coagulate with it instead of the little pencil instrument and the foot switch. Cushing looked at it and commented that it was very nice, and suggested that Bovie go and put a pencil in the muzzle of a pistol and write his name with it, and then he would see why surgeons rather than engineers should design surgical equipment.

CUSA in Neurosurgery

B.M. STEIN

Since the introduction of the ultrasonic aspirator developed and adapted for neurosurgery by the Cooper Scientific Corporation over the past decade, the instrumentation has gained wide acceptance in the removal of certain central nervous system tumors.

The tumors in which the use of this instrument is applicable are generally firm, or hard tumors located in obscure regions of the brain. Because there is limited hemostasis with this instrument, it has not been useful in soft tumors, especially those that are highly vascular (Table 1).

A number of examples of the tumors in which this has been successfully used will be reviewed and illustrated (Figs. 1–5) (Table 2).

Because the design of the instrument does not limit its bulk, it has been difficult to use the instrument effectively in some of the deeper located tumors. However, the development of a long curved tip has increased the usefulness of the instrument to some extent (Fig. 6). There are still problems with the bulk of the instrument and a disadvantage of the longer tip is that it extends the instrument further beyond the operation field sometimes interfering with free movement of the operation microscope.

In spite of these limitations and defects in the design, the instrument can be used along the base of the skull to remove meningiomas intertwined in the cranial nerves and indenting the brain stem; it has probably found its greatest usefulness in removal of acoustic tumors. In addition, it may be used in the pineal region with some difficulty (Fig. 4). Generally, it precludes the use of other instrumentation. In a similar fashion, it has been used within the ventricles for particularly firm or heavily calcified tumors (Fig. 5).

Table 1

Advantages
1. Rapid removal of tumor
2. Controlled removal of tumor
3. Minimal manipulation of structures surrounding tumor
4. Removal of firm tumors in obscure locations
Disadvantages
1. Bulky instrument
2. Not designed for microsurgery
3. Long curved tip poorly designed
4. Apparatus hard for nursing to set up

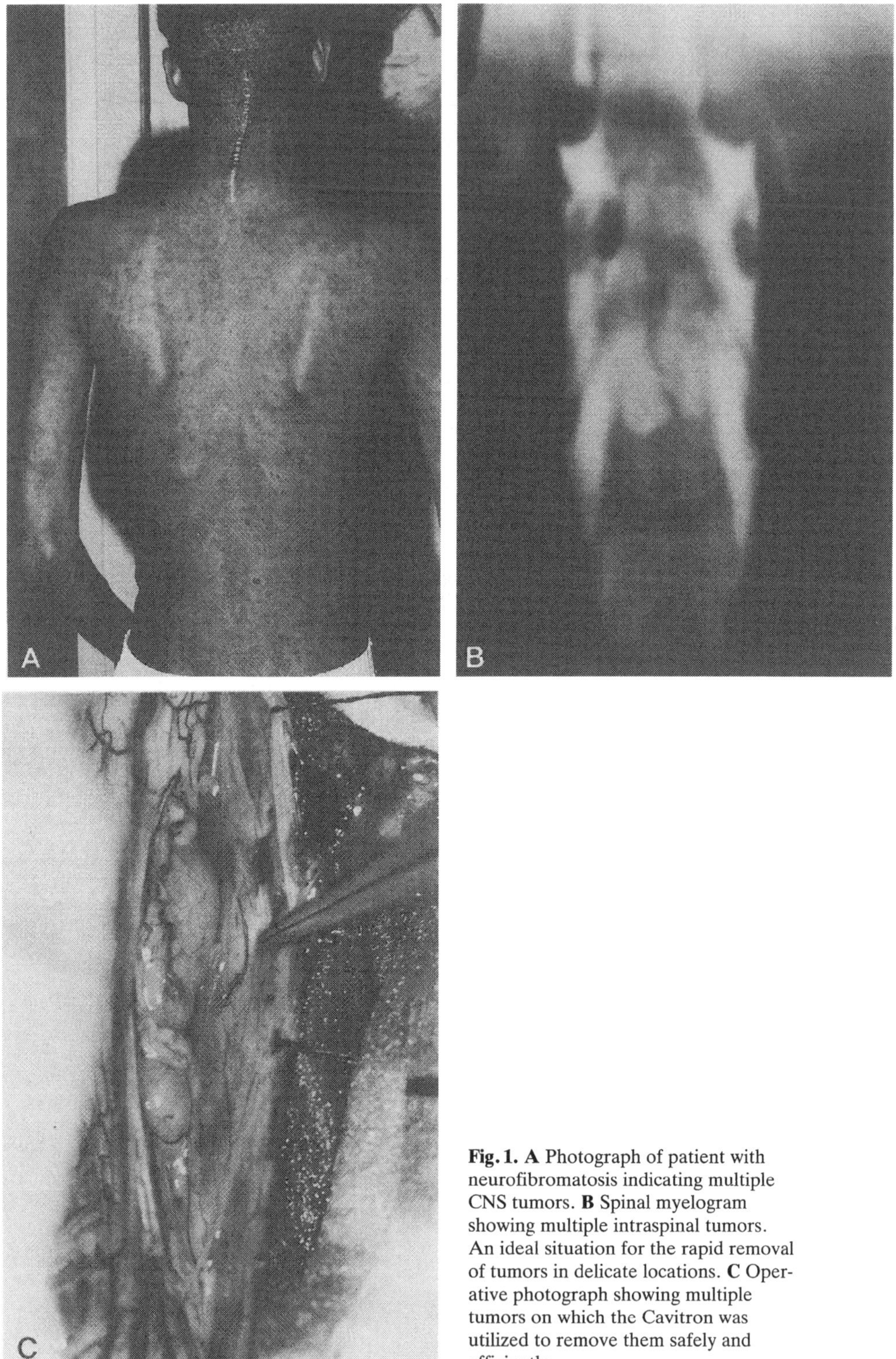

Fig. 1. A Photograph of patient with neurofibromatosis indicating multiple CNS tumors. **B** Spinal myelogram showing multiple intraspinal tumors. An ideal situation for the rapid removal of tumors in delicate locations. **C** Operative photograph showing multiple tumors on which the Cavitron was utilized to remove them safely and efficiently

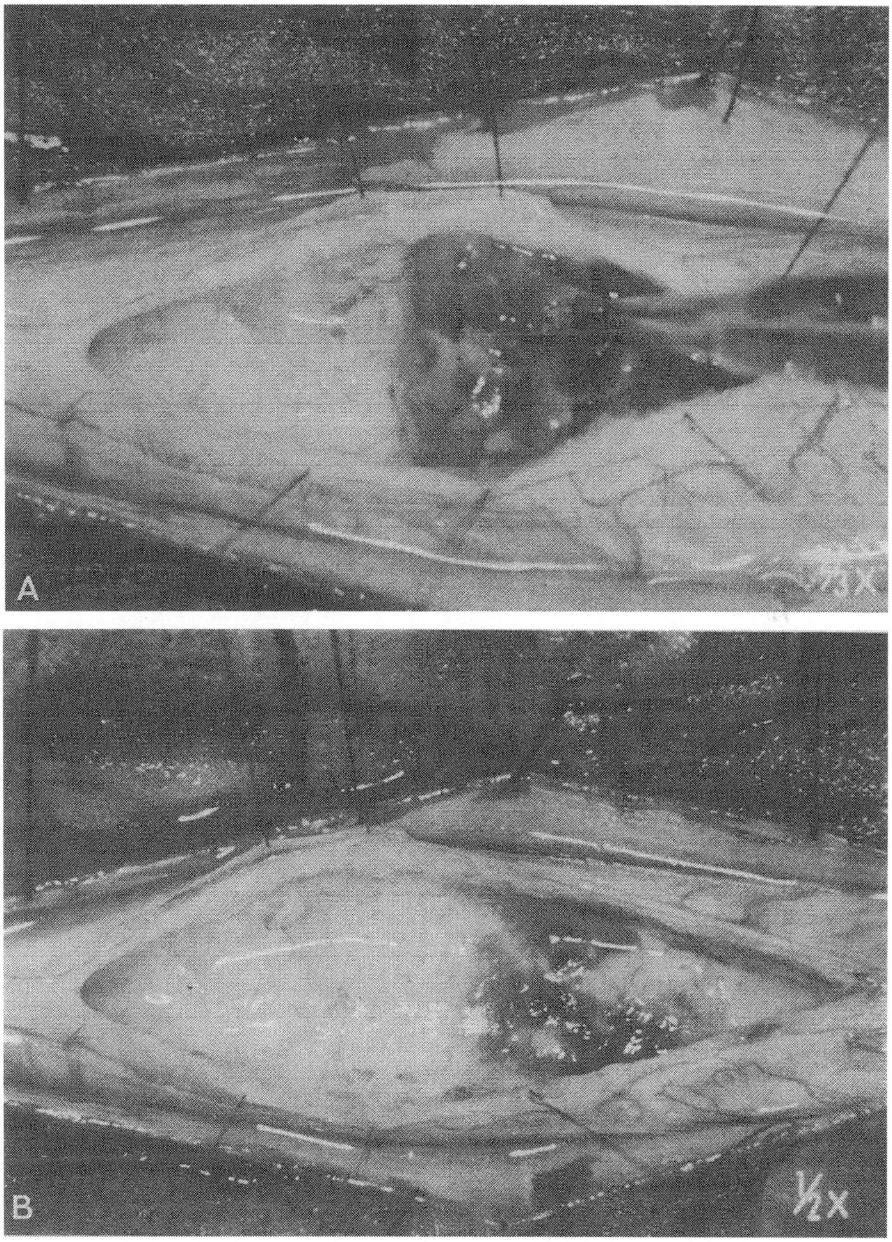

Fig. 2. A Photograph of intramedullary astrocytoma in a state of removal. **B** Intramedullary astrocytoma totally removed through the use of the Cavitron facilitating the removal and increasing the safety of the operation

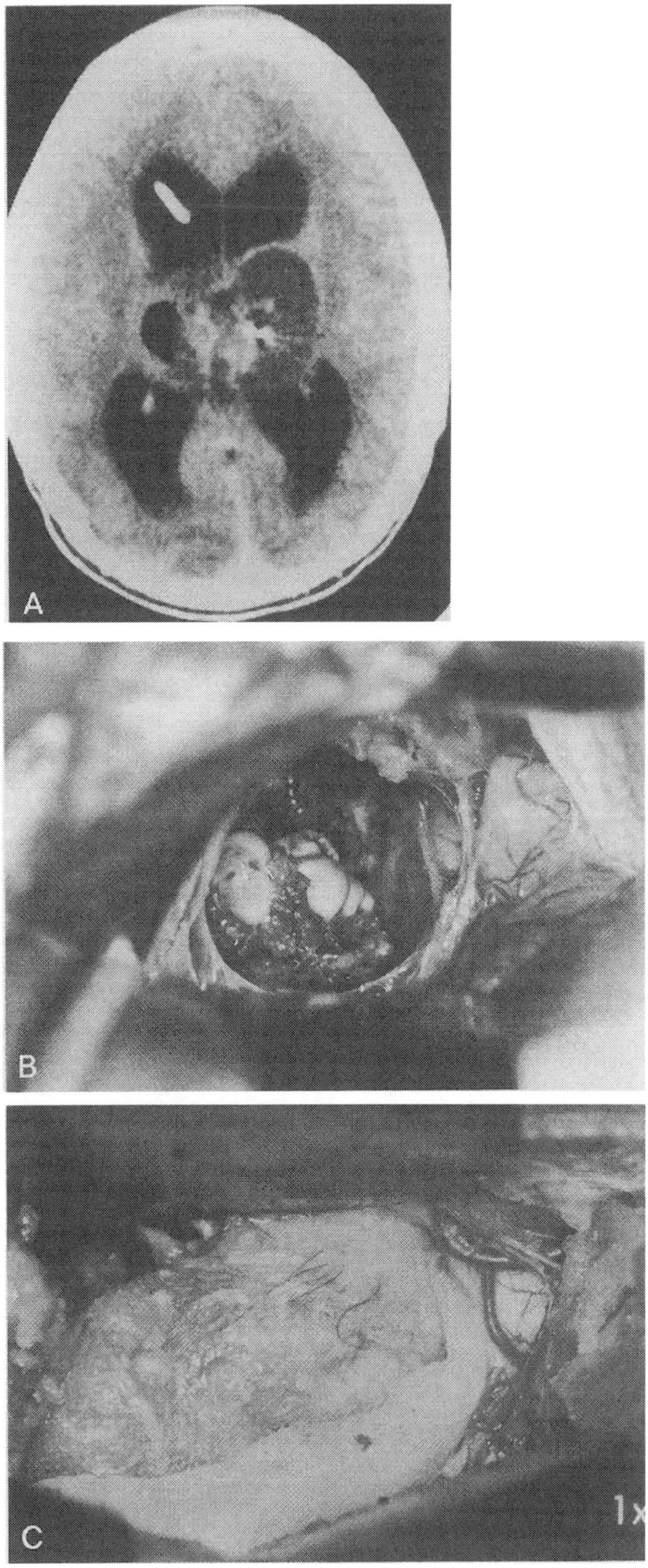
A
B
C
1x

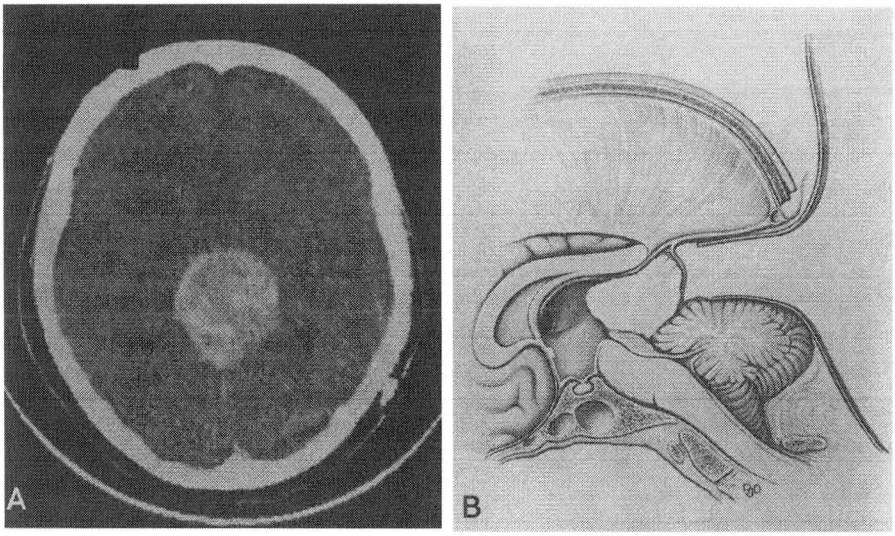

Fig. 4. A CT scan showing large meningioma of the pineal region. The Cavitron was used to remove this tumor. **B** Posterior fossa exposure used in the removal of this meningioma

Fig. 3. A CT scan showing large intraventricular tumor of a teratomatous nature. **B** Posterior fossa exposure of tumor indicating firm nature of component of tumor. **C** Second stage transcallosal exposure of large teratomatous tumor, removal facilitated by the Cavitron

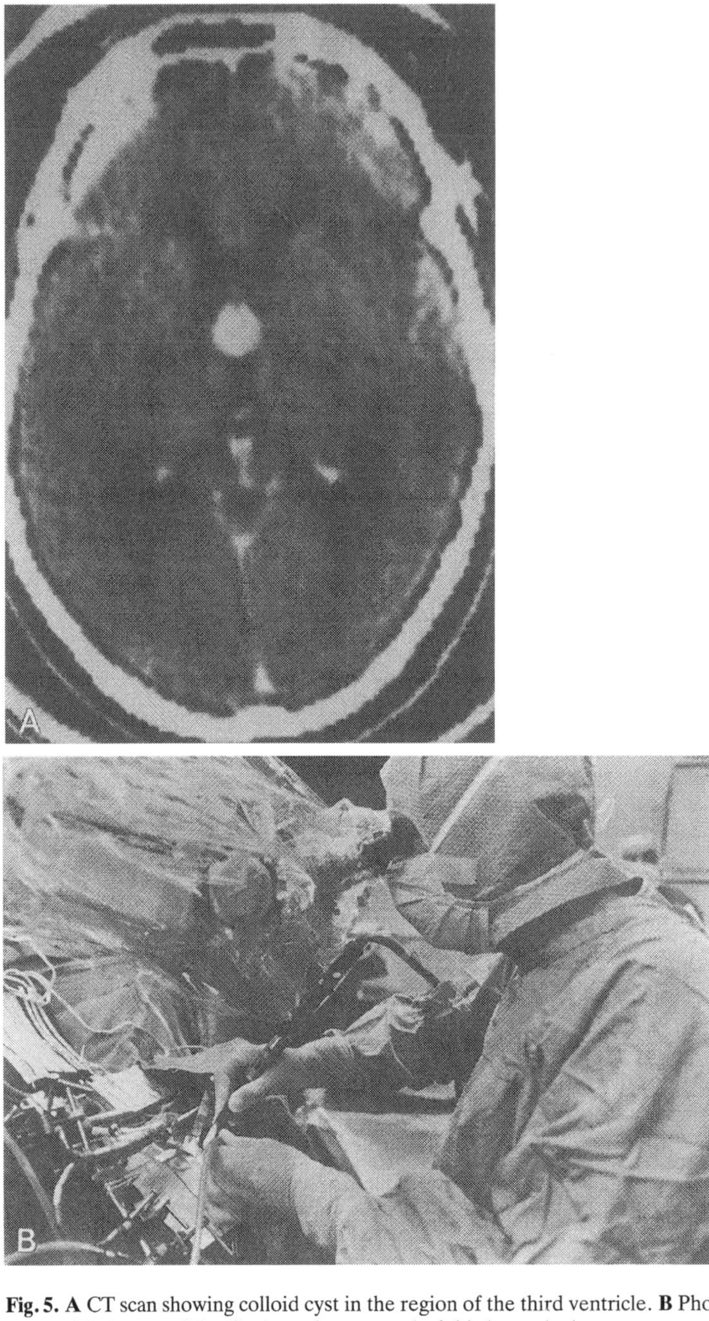

Fig. 5. A CT scan showing colloid cyst in the region of the third ventricle. **B** Photograph of operating room with the use of the Cavitron for removal of third ventricular tumor

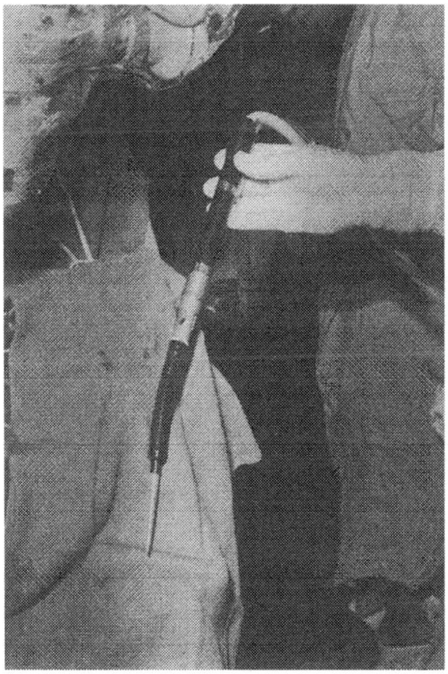

Fig. 6. Photograph of Cavitron hand piece showing the long curved tip

Table 2

Useful:
Basal meningiomas
Acoustic tumors
Intramedullary spinal tumors
Intraventricular firm tumors
Pineal tumors
Not useful:
Gliomas of brain
Medulloblastoma
Intraventricular soft tumors
Highly vascular tumors

Familiarity with the instrument allows the surgeon to adjust to various modalities to suit each individual case (Fig. 7). For example, on a highly calcified tumor, high flow irrigation and high vibration settings are essential. Contrariwise when pealing a moderately firm or soft tumor from the vessel, low vibration and irrigation setting will permit the surgeon to brush such tumors away from major vessels without injuring the vessel wall if it is done in a rapid and gentle fashion.

Unfortunately, the nursing staff is not able in most instances to familiarize itself with the instrument and often the set up requires a number of trials of the programming through the various phases that are involved in bringing the machine to full operating modality.

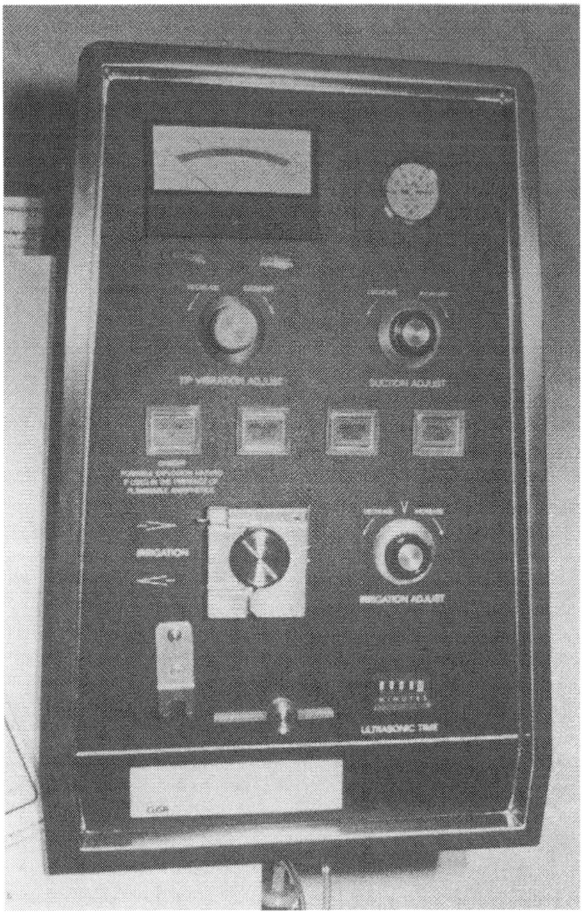

Fig. 7. Control panel of the Cavitron showing controls having vibration, suction and irrigation

It is hoped that the great advantages of this instrument in rapidly removing firm tumors without distorting or injuring surrounding structures will be further improved by design changes to miniaturize the instrument. A number of experimental studies have been tested for the safety factor of this instrument.

Spinal Cord Astrocytomas of Childhood

F. Epstein

Introduction

Intramedullary spinal cord astrocytomas are a relatively uncommon neoplasm, accounting for only 4% of central nervous system tumors of childhood (Anderson and Carson 1953; Areseni et al. 1967). Because of its rarity, individual neurosurgeons have relatively little experience with surgical management and long term follow-up of afflicted patients. There has been little impetus to modify the traditional treatment of biopsy, dural decompression and radiation therapy despite the recognition that after a relatively short remission serious disability or death ensues.

This must be regarded as particularly tragic as most of these neoplasms are low grade gliomas and microscopically identical to their "sister" tumors which occur in the cerebellum, and are surgically curable. It was this perspective that encouraged the author to explore the technical feasibility of radical excision of spinal cord astrocytomas. In this endeavour, 120 children have undergone gross total excision of spinal cord astrocytomas over the past 4 years. This unusual series has provided a unique opportunity to study the biology of the tumor as well as its response to traditional, as well as recently introduced more radical surgery techniques.

Cervical Tumors

The most common early symptoms were nuchal pain and head tilt with torticollis. Mild upper extremity monoparesis was the next most common symptom, and was often extremely subtle during the early stages of the illness. Very often, in young children, the first manifestation of weakness was switching "handedness" in right- or left-handed patients. Neoplasms in the caudal cervical spinal cord commonly caused weakness and atrophy of the intrinsic muscles of the hand in contradistinction of tumors rostral to C5 which were less likely to cause significant weakness until relatively late in the clinical course. Interestingly, weakness of the lower extremities only evolved months or, rarely, 2 to 3 years after the first symptoms, and bowel and bladder dysfunction was rarely present at the time of primary diagnosis.

Sensory abnormalities were generally limited to one upper extremity, and a discrete sensory level was only noted very late in the course of the disease and, then, only in association with severe neurological disability.

In most patients, there was increased reflex activity in the lower extremities with or without extensor plantar signs and clonus.

Thoracic Tumors

Mild scoliosis was the most common early signs of an intramedullary thoracic cord neoplasm. Pain and paraspinal muscle spasm commonly occurred before there were objective signs of neurological dysfunction, and were commonly assumed to be secondary to the evolving scoliosis. Insidious progressive motor weakness in the lower extremities was first manifest by "awkwardness" and, only later, by frequent falls and an obvious limp. Early sensory abnormalities were common although dysesthesias and paresthesias were occasionally present. Increased reflexes and extensor plantar signs, with or without clonus, occurred relatively early in the neurological course.

A presenting complaint of bowel and bladder dysfunction was most unusual, and was diagnostic of neoplasm extending into the conus. In general, these symptoms evolved only late in the clinical course if the tumor was rostral to the conus medullaris.

Neurodiagnostic Studies

Spinal cord astrocytomas may be divided into two general categories: holocord and focal.

Holocord Astrocytomas

"Holocord" widening occurred in 60% of pediatric patients, and was manifest by expansion of the entire spinal cord from the medulla or cervicomedullary junction to the conus.

These neoplasms were invariably cystic astrocytomas in which the solid component of the neoplasm spanned a variable length of the cord, and was associated with huge non-neoplastic rostral and caudal cysts which expanded the central canal above and below the tumor.

Plane spine X-rays commonly disclosed a diffusely widened spinal canal with relatively localized erosion or flattening of pedicles. Whereas the former was secondary to long-standing expansion of the entire spinal cord, the latter occurred only adjacent to the solid component of the neoplasm.

Although there were occasional early case reports of holocord widening, its relative frequency was probably not recognized because of the tendency to terminate the neurodiagnostic study when a lumbar Myelogram disclosed a complete block secondary to an intramedullary neoplasm (Areseni et al. 1967; Coxe 1961). In the first patient in this series a cervical puncture was employed to identify the rostral extent of cord widening. It was subsequently recognized that although not apparent on the myelogram, a small amount of Metrizamide almost invariably "trickles" past the block and is obvious on the immediate or delayed spinal CAT scan which, therefore, defines the rostral extent of the expanded cord (Epstein 1984a; Epstein and Epstein 1981; 1982a; 1982b). It is for this reason that the availability of Computerized Axial

Tomography (CAT) of the spine is an invaluable adjunct to the neurodiagnostic evaluation of spinal cord tumors.

A 24-hour delayed spinal CAT scan was a mandatory study as rostral and caudal, as well as occasional intra-tumor cysts were identified as the Metrizamide diffused within them.

Transcutaneous Ultrasound

In patients who have had an earlier laminectomy, transcutaneous ultrasonography may be employed to visualize the spinal cord and the neoplasm. Utilizing this technique (real-time unit with triple frequency: 3.5 and 7.5 Mhz) the cord may be studied in both sagittal and transverse projections, and the presence of significant expansion will be immediately obvious (Braun et al. 1983; Epstein and Epstein 1983; Raghavendra et al. 1984).

Transcutaneous ultrasonography may be more informative than conventional myelography or Metrizamide spine CT scanning as it gives a direct view of the interior of the spinal cord. In occasional cases, the tumor may be echogenic, affording a dramatic view of the neoplasm and its relationship to the spinal cord. The presence of cysts either within the tumor or in relation to the rostral and caudal pole of the neoplasm is evident. 18 months or longer, following radiation therapy, there are commonly multiple intra-tumor cysts which are "swiss cheese" like in appearance. The technique is limited by the length of the laminectomy, and it is not possible to visualize the spinal cord rostral or caudal to the laminectomy defect.

Surgery

It is desirable to carry out a limited laminectomy over the solid component of the neoplasm, but not to unnecessarily extend it rostrally or caudally.

In our first surgical experience with "holocord" widening, a total laminectomy form C1-T12 was carried out. It was subsequently recognized that it was not necessary to expose the spinal cord over the rostral and caudal cysts and, for this reason, it was important to define as accurately as possible this location of the solid component of the neoplasm vis-a-vis the cysts.

The approximate location of the solid component of the neoplasm may be estimated on the basis of the clinical and radiographic findings.

Clinical Indications of Tumor Location in the Presence of Holocord Expansion

In the presence of holocord widening associated with a cystic astrocytoma, it is the solid component of the neoplasm that is responsible for primary neurological dysfunction while the rostral and caudal cysts which expand the remainder of the spinal cord remain asymptomatic in the early stages of the disease.

Therefore, neurological symptoms in one or both upper extremities in the presence of holocord widening suggests that the solid component of the neoplasm is within the cervical cord.

Conversely, progressive scoliosis and/or neurological dysfunction limited to the lower extremities are strongly suggestive of solid neoplasm within the thoracic cord, while bowel and bladder dysfunction indicates extension of neoplasm into the conus.

In the presence of normal bowel and bladder function, an expanded conus, in our experience, is invariably associated with a cyst.

Spinal cord ependymomas do not adhere to this clinical pattern as they may expand any length of the spinal cord with a relative paucity of signs and symptoms referrable to the segmental involvement. It is tempting to speculate that this is directly related to the primary anatomic location of the tumor in the region of the central canal which causes very gradual compression of adjacent neural structures as the tumor increases in volume. This may be analogous to the rostral and caudal.

Surgical Technique

In patients who have not been previously operated on, an osteoplastic laminectomy is carried out. This permits replacement of the bone which is a nidus for subsequent osteogenesis and posterior fusion. Replacement of the bone does not prevent the post-surgical evolution of spinal deformity, but offers protection against future local trauma.

Even following careful consideration of the clinical and neuroradiologic examination, it is not possible to be certain that the laminectomy is of sufficient length to expose the entirety of the solid component of the neoplasm. For this reason, transdural ultrasonography is utilized to further define the location of the tumor vis-à-vis the bone removal.

Therefore, after laminectomy is carried out, the wound is filled with saline and the head of the transducer probe is placed into gentle contact with the dura. Utilizing this technique, the spinal cord is viewed in both sagittal and transverse sections. The rostral and caudal limits of the tumor, as well as the presence or absence of associated cysts are immediately obvious (Fig. 1). Occasionally, an echogenic tumor provides a striking ultrasound image though, most commonly, the solid component of the neoplasm is only manifest as a widened spinal cord. If the laminectomy is not sufficiently long to expose the entirety of the solid component of the neoplasm it is lengthened, segment-by-segment, until the ultrasound discloses that the entire tumor mass is exposed.

Only at this juncture is the dura opened, and this is limited to overlay the expanded spinal cord – it is not extended rostrally or caudally over normal spinal cord. In addition, it is not necessary to open the dura widely over the rostral or caudal cyst, as these are easily drained as the solid component of the neoplasm is excised.

It is rarely helpful to utilize intra-operative ultrasound after the dura is opened as numerous surgical artifacts obscure the image.

It is important to emphasize that the swollen spinal cord is commonly rotated and distorted, and it is essential that careful inspection identify normal landmarks prior

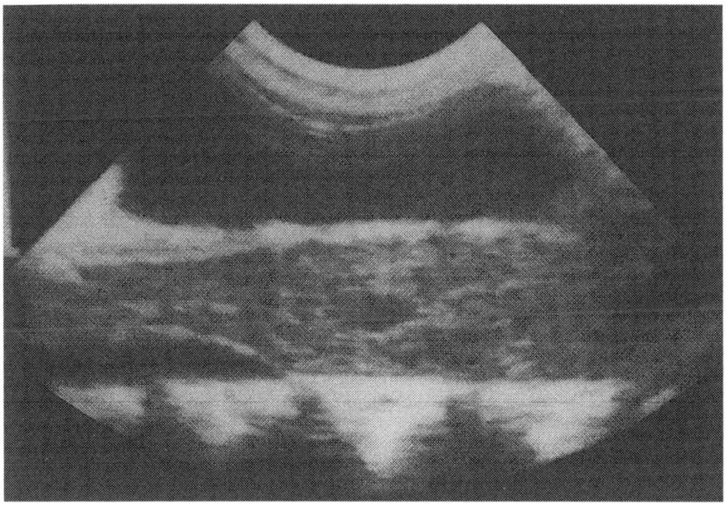

Fig. 1. Intraoperative ultrasonography discloses sagittal view of normal and expanded spinal cord. Note presence of small intra-tumor cyst

to placing the myelotomy. Since the posterior median raphe is generally obliterated, the only sure way of recognizing the posterior midline is by identifying the dorsal root entry zones bilaterally. Rotation of the cord may occasionally make this difficulty, and even surprising, in terms of the distorted location of the midline. In any event, this is important as otherwise it is possible that the myelotomy may be placed away from the median raphe and sever multiple nerves along the dorsal root entry zone.

In the presence of holocord widening associated with rostral and caudal cysts, the ultrasound will have clearly defined the junction of cyst and neoplasm over the rostral and caudal poles of the tumor. It is in the "junctional" regions that the midline myelotomy is started.

The carbon dioxide laser utilized at 6–8 watts is an ideal instrument for placing the myelotomy as the cord is incised and hemostasis obtained simultaneously. Although neurosurgeons are loath to interrupt blood vessel on the surface of the spinal cord, it is tedious and time-consuming to preserve these vascular channels, not at all essential to the preservation of neurological function, and they are almost inevitably disrupted during the course of the procedure, even if primarily preserved.

After the cyst is entered, inspection of the cavity will localize the rostral or caudal neoplasm which extends into it. It is not necessary, in most cases, to extend the myelotomy over the cyst as it is easily drained as either pole of the neoplasm is identified and removed. Because the cyst fluid is produced by the tumor, it is unlikely to re-accumulate following gross total excision of the neoplasm.

After identifying the rostral and caudal cyst-tumor junction, the myelotomy is continued over the midline of the cord between the previously placed incisions.

Following completion of the myelotomy, there is usually 1 to 2 mm of white matter overlying the neoplasm which is removed with the laser or bipolar cautery and a

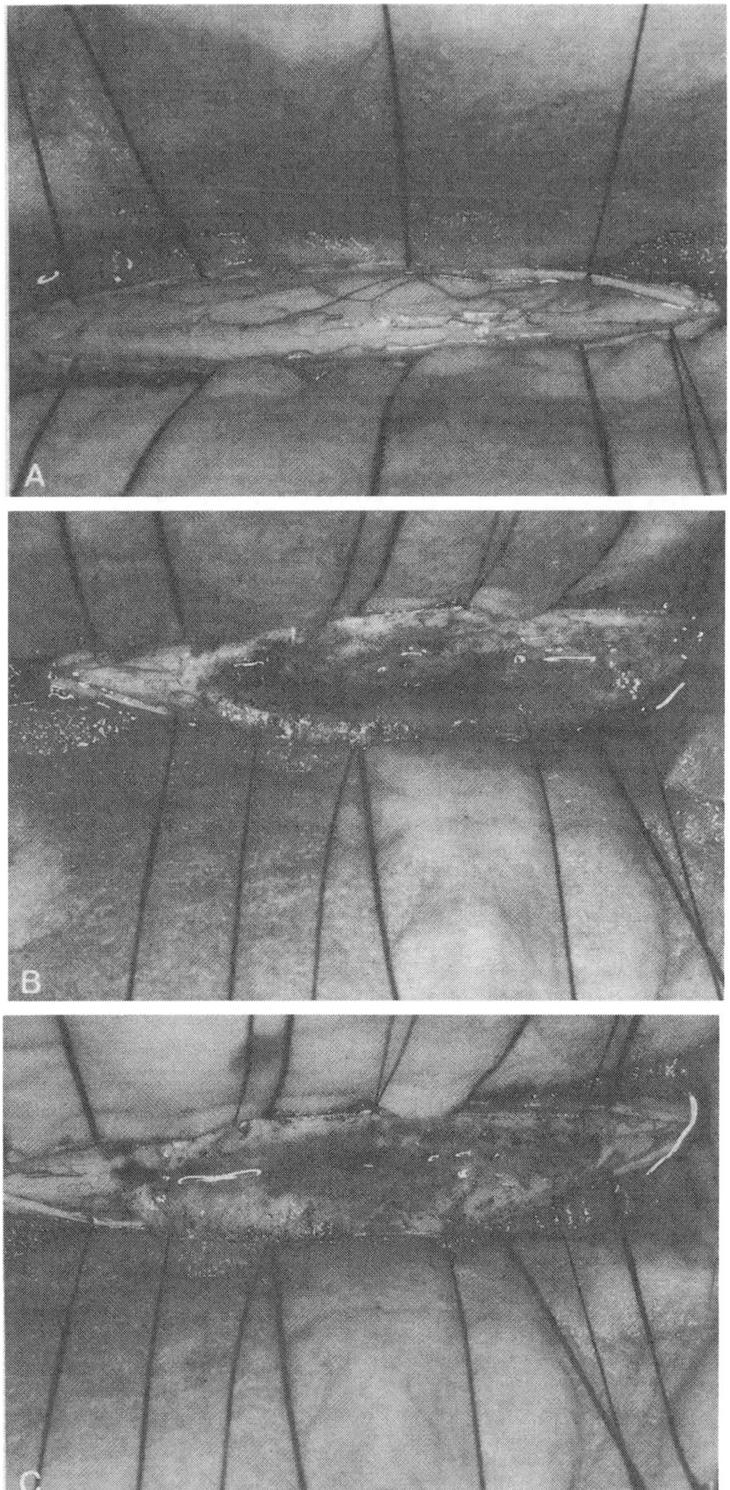
A
B
C

very fine suction. Most astrocytomas are grey or pink in color, and may be distinguished from adjacent white matter.

At this juncture, it is essential that pia traction sutures be utilized to open the myelotomy incision and further expose the intramedullary tumor. It is satisfactory to utilize any fine suture material, and it is the author's practice to simple "hang" small clamps on the sutures rather than suturing the pia to adjacent tissues.

It must be emphasized that in the presence of an astrocytoma, there must be no effort to define a plane of cleavage around the tumor. These neoplasms must be removed from "inside out" until a "glia-tumor interface" is recognized by the change in color and consistency of the adjacent tissues. There is rarely a true plane of dessection, and futile efforts to define its presence only results in unnecessary retraction and manipulation of functioning neural tissue.

In the presence of cystic holocord neoplasm, tumor removal is initiated either at the rostral or caudal pole of the neoplasm in the region of the tumor-cyst junction.

As tumor excision continues, it is helpful to recognize that the anterior extent of the neoplasm is only very rarely ventral to the anterior wall of the cyst. The bulk of the tumor is most often in the posterior two-thirds of the spinal cord (as viewed in cross-section), and the general dimensions of the tumor may be roughly conceptualized following inspection of the rostral and caudal cyst.

The excision of the solid non-cystic astrocytoma is initiated in the mid-portion rather than the rostral or caudal pole of the neoplasm. This is because there is no clear rostral or caudal demarcation of the tumor as occurs when there are rostral and caudal cysts. In addition, the "poles" of the neoplasm are the least voluminous and, for this reason, removal of this part of the neoplasm may be the most hazardous as normal neural tissue may be easily disrupted.

The CUSA is utilized to remove the bulk of the neoplasm, following which the carbon dioxide laser is employed to vaporize the visible remaining fragments (Fig. 2A–C). The dura is closed primarily as it is unnecessary to utilize a dural substitute for decompression if tumor excision has been grossly complete.

Discussion

There are a number of important observations that are clearly relevant in terms of understanding the biology of this group of neoplasms, as well as recommending proper surgical management. It has been a consistent observation that in the presence of holocord expansion, the solid component of the astrocytoma is often not as extensive as myelography alone suggests and, indeed, the actual location of the neoplasm may be in those segments of the spinal cord that correspond to neurological dysfunction. The lack of significant neurological dysfunction relating to spinal segments that were distended with fluid is probably directly related to the anatomical

Fig. 2. A External appearance of expanded spinal cord. **B** Following placement of myelotomy and pia traction sutures the tumor is visualized. **C** After "gross total" excision of tumor, there is large intramedullary cavity. This patient made a complete neurological recovery location of the cyst within the center of the cord as compared to the solid component of the neoplasm which was relatively diffuse.

The presence of cysts which similar in appearance to those associated with the cystic astrocytoma of the cerebellum suggests that the neoplasms are congenital tumors that have their inception sometime during gestation. The fluid produced by the tumor extends up and down the spinal cord in the region of least resistance, that is, the central canal.

One might also speculate that in some cases the classical symptoms of syringomyelia may, in fact, be a late manifestation of such a cyst in which the tumor has either involuted or is not anatomically obvious. Perhaps the centrally located cyst may gradually expand over many years and compress the surrounding cord. In this regard, it is significant that a few patients with holocord widening had exceedingly small neoplasms, between 1.5 and 3 cm, and were mistakenly diagnosed as syringomyelia or hydromelia. Our experience would suggests that the presence of xanthochronic cyst fluid is pathognomonic of an associated neoplasm, while clear fluid is diagnostic of hydromelia.

It is our perspective that the presence of a widened spinal cord from the cervicomedullary junction to the conus, which is associated with a relatively slowly evolving neurological deficit, is indicative of a very slowly growing and perhaps even haemartomatous-type of lesion which has a good long-term prognosis and should be treated aggressively.

Nevertheless, it must be emphasized that despite a "gross" total tumor excision, it would be naive to assume that residual tumor fragments were not commonly left in situ. We have hypothesized that these remaining fragments may remain dormant, or involute, in a similar way to what has been noted to occur in many astrocytomas of the cerebellum (Bucy and Theiman 1968; Geissinger and Bucy 1971). However, whether or not this is reality or "wish-fulfillment" will only be known many years from now following long-term follow-up and retrospective analysis.

In most cases of holocord tumor, the initial complaint was a weak arm, or a mildly weak leg, and associated pain somewhere along the spinal axis. The signs and symptoms were consistently relatively minor when compared to the apparently diffuse nature of the pathological process. It is perfectly understandable why neurosurgeons faced with this clinical dilemma have been most concerned about inflicting a great neurological deficit as a result of extensive dissection within a rather well-functioning spinal cord. This rationale has been used for a temporizing surgical approach consisting of a limited laminectomy and biopsy, and relying on radiation therapy to control tumor growth. Unfortunately, the natural history of these tumors, with radiation therapy, is slow deterioration and eventual severe neurological disability or death.

The outcome following radical resection of these tumors was directly related to the pre-operative neurological status. Although a transient increase in weakness or sensory loss was commonly present in the immediate post-operative period, only one patient had a significant permanent increase in neurological deficit following operation. Patients with paraparesis or quadriparesis, who were ambulatory before surgery had neurological improvement over several weeks. The group with severe deficits pre-operatively, rarely made any significant improvement although their downhill course abated.

There is no evidence that radiation will cure benign astrocytomas of the spinal cord, and there is abundant evidence that it has a deleterious effect on the immature developing nervous system. Spinal cord astrocytomas should be recognized as potentially excisable lesions, with radiation therapy reserved for possible adjunctive use if there is a recurrence. At that time, it might be employed following a second radical surgical resection.

Children who have undergone extensive laminectomy and, in addition, have denervation of the paravertebral muscles from tumor infiltration of anterior horn cells as well as operative muscle retraction, are at risk for developing severe spinal deformities as they pass through periods of rapid growth. Close collaboration with a pediatric orthopedic surgeon experienced with kyphoscoliosis is essential in following these patients.

Summary

The author has carried out gross total excision of an intramedullary spinal cord astrocytoma in 120 consecutive patients. This experience has led to the following conclusions:

1. Holocord widening occurs in 60% of cases, and is diagnostic of a cystic astrocytoma.
2. Despite the absence of a surgical plane of dissection, these neoplasms may be removed from "inside out" until a glia-tumor interface is recognized.
3. Radical tumor excision is compatible with partial or total recovery of neurological function.
4. The success of surgery is directly related to the pre-operative neurological status of the patient. Paralysis or near paralysis was never improved, while mild to moderate pre-operative neurological dysfunction often recovered.
5. While this experience has established the efficacy of radical surgery, there is no information to suggest the duration of remission, or the likelihood of permanent cure. This will only become known at the time of retrospective analysis many years from now.

References

Anderson FM, Carson MU (1953) Spinal cord tumors in children. A review of the subject and representation of twenty-one cases. J Pediatr 43 : 190–207

Areseni C, Horvath L, Illiescu D (1967) Intraspinal tumors in children. Psychiatr Neurol Neurochir 70 : 123–133

Braun IF, Raghavendra BN, Kricheff II (1983) Spinal cord imaging using real-time high-resolution ultrasound. Radiology 147 : 459–465

Bucy PC, Theiman PW (1968) Astrocytomas of the cerebellum. A study of a series of patients operated on over 28 years ago. Arch Neurol 18 : 14–19

Coxe WS (1961) Tumors of the spinal canal in children. Am Surg 27 : 62–73

Epstein F (1984a) The cavitron ultrasonic aspirator in tumor surgery. Clin Neurosurg 31:497–505
Epstein F (1984b) Radical surgical excision of spinal cord tumors of childhood. Technical considerations. Excerpta Medica
Epstein F, Epstein N (1981) Surgical management of "holocord" intramedullary spinal cord astrocytomas in children. J Neurosurg 54:829–932
Epstein F, Epstein N (1982a) In: McLaurin, Robert et al. (eds) Pediatric Neurosurgery: surgery of the developing nervous system. New York, Grune & Stratton, pp 529–539
Epstein F, Epstein N (1982b) Surgical management of extensive intramedullary spinal cord astrocytomas in children. Concepts Pediatr Neurosurg 2:29–44
Epstein F, Epstein N (1983) Surgical treatment of spinal cord astrocytomas of childhood. A series of 19 patients. J Neurosurg 75:685–689
Epstein F, Raghavendra NB, John RE, Pritchett L (1985) Spinal cord astrocytomas childhood: surgical adjuncts and pitfalls. Concepts Pediatr Neurosurg 5:224–237
Flamm ES, Ransohoff J, Wachinich D, Broadwin A (1978) A preliminary experience with ultrasonic aspiration in neurosurgery. Neurosurgery 2:240–243
Geissinger JD, Bucy PC (1971) Astrocytomas of the cerebellum in children. Long term study. Arch Neurol 24:125–135
Raghavendra BN, Epstein FJ, McCleary L (1984) Intramedullary spinal cord tumors in children: localization by intraoperative sonography. Amer J Neuroradiol 5:395–397
Schijman E, Zuccaro G, Monges JA (1981) Spinal tumors and hydrocephalus. Child Brain 8:401–405

Treatment of Arteriovenous Malformations of the Brain and Spinal Cord

B.M. Stein

Brain AVM

Introduction

Personal experience in the treatment of 250 arteriovenous malformations of the brain is presented. 180 of these patients underwent surgery, with a mortality of 1.5%. A morbidity rate of 10% includes cases of hemianopsia after occipital malformations had been removed. Significant morbidity of aphasia, hemiparesis and hemisensory loss, occurred in under 5% of cases.

Over the past decade, I have personally treated 250 patients with arteriovenous malformations (AVMs) of the brain. 180 of these patients underwent surgery and it is on this group that this presentation is focused. The other patients in the series were treated primarily by embolization (Wolpert and Stein 1975; 1979). Many of the surgical cases had embolization and this will be discussed when it applies as an adjuvant to surgery.

Pathophysiology

AVMs of the brain are considered congenital disorders which manifest, in the vast majority of instances, by hemorrhage and to a lesser extent by seizures in the middle decades of a patient's life (Michelsen 1979; Morello and Borghi 1973; Pool and Potts 1965; Stein 1979; Stein and Wolpert 1980a; 1980b; Troupp 1976). Rarely is the first hemorrhage fatal, but it may leave the patient with a significant neurological deficit (Fig. 1, 2). Recurrent hemorrhages often lead to catastrophic neurological disability or death. It is for this reason that AVMs are considered dangerous to the patients' well-being, especially in a young individual in the prime of a productive life. It is also generally accepted that surgical removal is the only way to insure that the malformation will not cause further difficulties. Anything less than total obliteration of the AVM, such as clipping of arterial feeders or embolization of portions of the malformation, is generally considered a less than complete form of therapy (Fig. 3).

The majority of AVMs occur in the supratentorial region over the convexity of the brain. Approximately 10% are located in the posterior fossa, either in the brain stem or the cerebellum. A lesser number of malformations involve deep portions of the cerebral hemisphere (Garrido and Stein 1978; Juhasz 1978). These would include intraventricular and diencephalic malformations. Also, less common than convexity

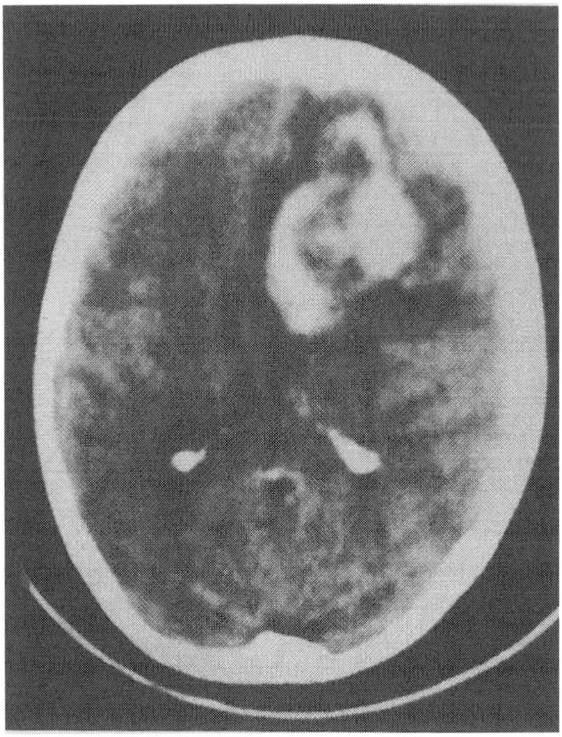

Fig. 1. CT scan showing large intracerebral hemorrhage from an AVM

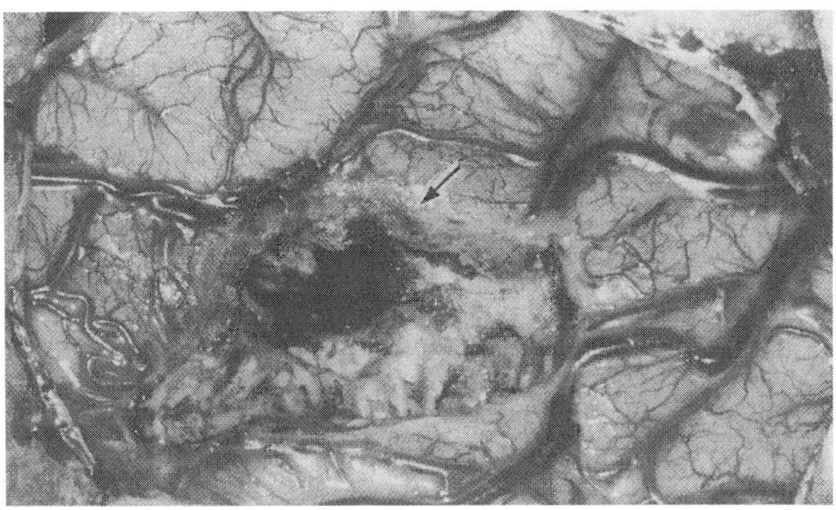

Fig. 2. Operative photograph of same case showing large hematoma *(arrow)* deep to the resected AVM. This patient suffered irrevocable neurological deficit from the hemorrhage

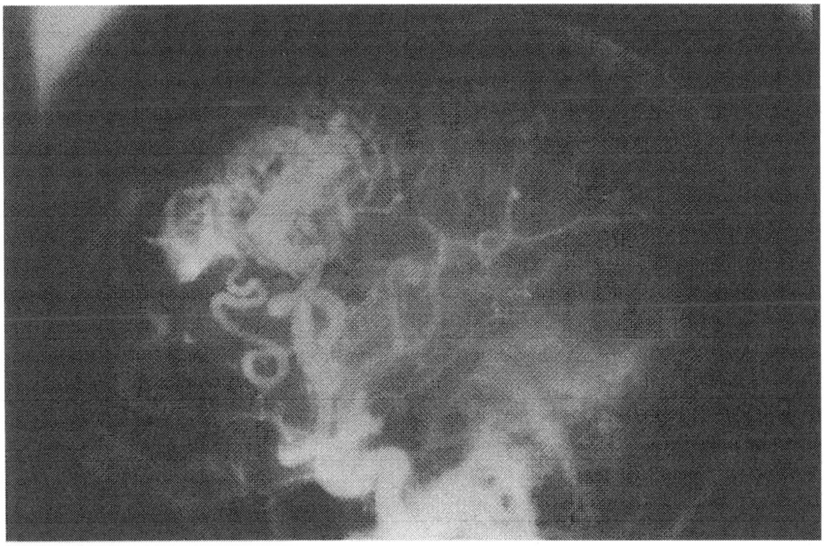

Fig. 3. Lateral carotid arteriogram showing a large left frontal AVM. This patient received proton beam therapy in a futile attempt to treat the malformation

AVMs are those located on the medial side of the cerebral hemispheres which are accessible but can be difficult to resect (Yasargil et al. 1976).

The data from my series would corroborate the above statistics. The results of AVM resection have improved with experience. Approximately 25% of malformations are inoperable (Kunc 1974); these are large and involve the major portion of a cerebral hemisphere or are located deep and involve extensive regions of the ventricular and diencephalic areas.

Before the operation the patient and his family must be psychologically prepared for the risks inherent in this major form of neurosurgery. This requires repeated and long discussions concerning the pros and cons of operating. The risk of surgical obliteration of an AVM are primarily related to factors such as location, size, age of the patient and the surgeon's experience with AVMs (Amacher et al. 1972; Drake 1979; Guidetti and Delitala 1980; Wilson and Dominque 1979).

Preoperative Embolization

I have found preoperative embolization a helpful adjuvant to the surgical resection of some of the more complex AVMs (Fig. 4). (Luessenhop and Presper 1975; Stein and Wolpert 1977). This has been most useful in the following circumstances: 1) for a malformation with a deep arterial feeder which is accessible to embolization, i.e., those which are fed by a significant posterior cerebral component which may be embolized effectively; 2) to reduce the turgor and extent of a malformation prior to surgical intervention; 3) to define more precisely the margins of an AVM by occluding those portions that extend beyond the mass of the malformation; and 4) to prepare

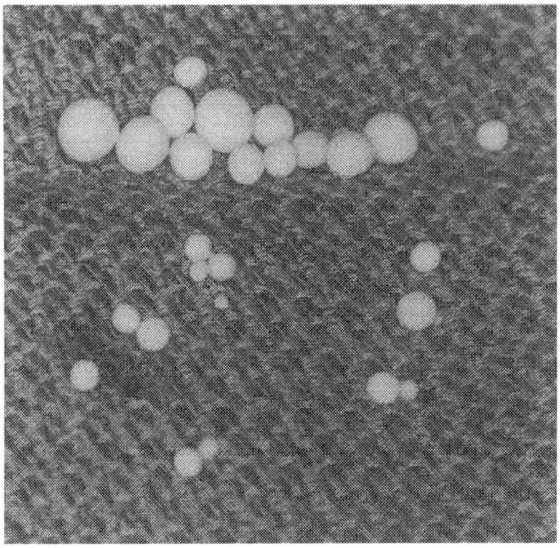

Fig. 4. Silastic pellets of various size used in embolization

critical cerebral arteries for eventual occlusion when they supply both the malformation and the normal brain.

With new intravascular microballoon techniques, the radiologists are now able to direct the flow of silastic pellets within the blood stream. Heretofore, pellets followed the anatomical course of the blood flow and it had been impossible to direct them at right angles to the flow or at other acute angles. However, now a microballoon catheter may be used to protect crucial arteries, therefore redirecting pellets into deep arterial feeders which may come off in a perpendicular direction to the mainstream. The following case is illustrative: in Fig. 5, a large arteriovenous malformation of the right sylvian region is shown to be supplied by major branches of the middle cerebral artery in the sylvian fissure. In addition, there is a large lenticulostriate feeding artery. Pellets were initially placed in the peripheral feeders to this AVM through standard techniques. Then a balloon catheter was placed in the middle cerebral artery just distal to the large lenticulostriate feeder and temporarily used to occlude and protect the main stem of the middle cerebral. Emboli were then injected and redirected into the lenticulostriate artery occluding it.

Besides silastic embolization, other techniques using particulate matter or liquids which solidify upon reaching the blood stream have been utilized. However, these other techniques carry a greater risk while they may be more effective. We have found for pratical purposes that silastic embolization has carried a good range of safety in the neighborhood of 1% to 2% risk while the injection of liquids or the use of macrointravascular balloon catheters has considerably increased the risk of the procedure.

The Role of Arteriography

Quality arteriography is absolutely essential in planning the operation and assessing the risks involved. I prefer to have stereoscopic lateral views which give a better con-

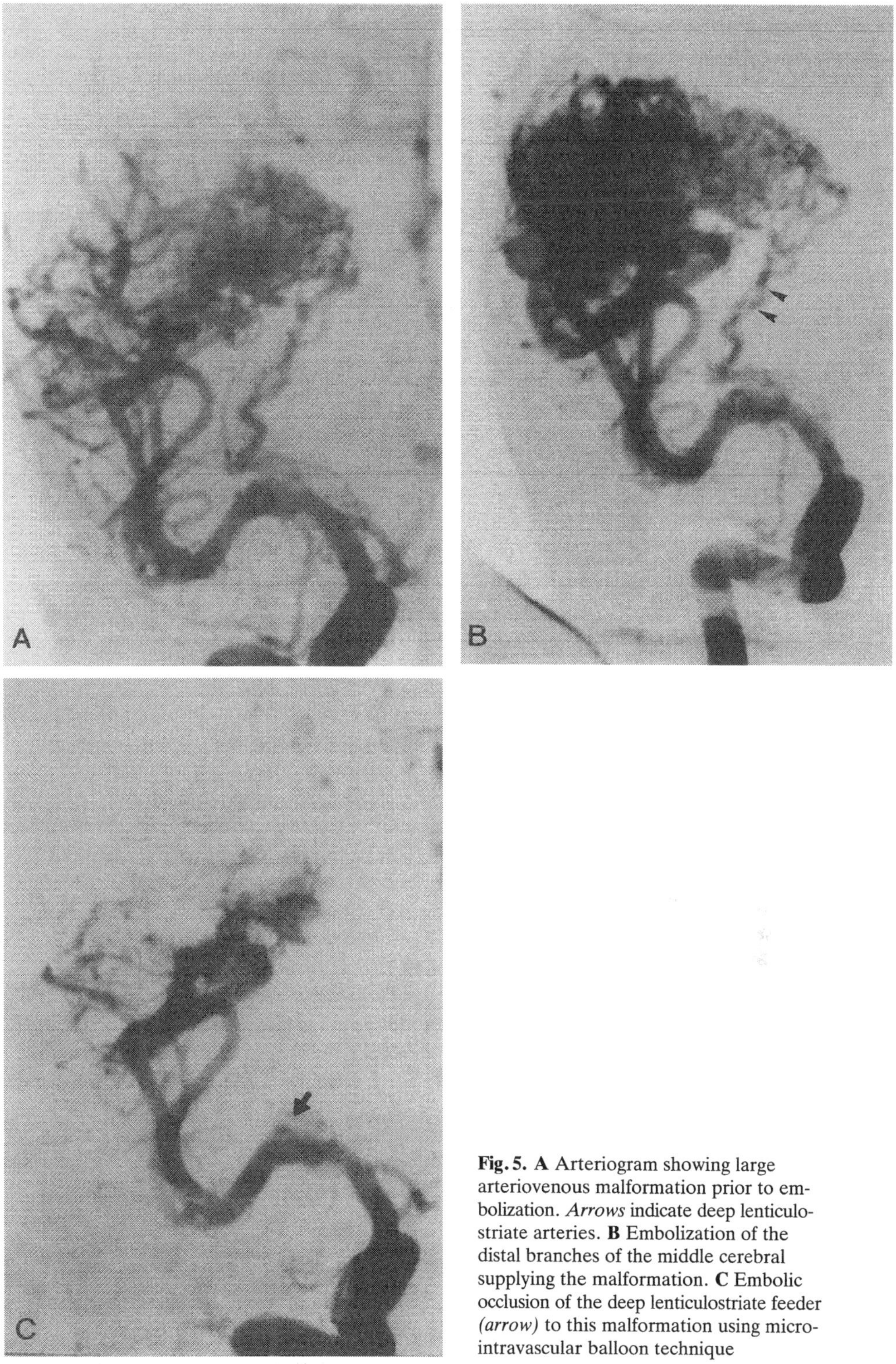

Fig. 5. A Arteriogram showing large arteriovenous malformation prior to embolization. *Arrows* indicate deep lenticulostriate arteries. **B** Embolization of the distal branches of the middle cerebral supplying the malformation. **C** Embolic occlusion of the deep lenticulostriate feeder *(arrow)* to this malformation using micro-intravascular balloon technique

cept of the three-dimensional anatomy of the vascular supply and venous drainage of the AVM. Computed tomography (CT) with and without contrast is useful in determining the relationships, especially of deep portions of the AVM, to associated structures such as the ventricular system, the diencephalon and other subcortical structures. Furthermore, CT will indicate the extent of a hematoma when the AVM has caused hemorrhage. Arteriographic findings that alert the surgeon to difficulty in resection of the malformation include the following: (1) Large veins from the AVM draining into the ependymal venous system, (2) blood supply from the choroid plexus of the lateral or third ventricles, (3) major deep arterial feeders from the perforating arteries such as lenticulostriate arteries, (4) poorly defined borders to the AVM, especially in those malformations which are located in or close to eloquent areas of the brain such as motor, speech and sensory regions and, (5) the association of single or multiple aneurysms with an AVM. In the latter instance, although controversy exists, it is my feeling that the aneurysm is the more likely cause of bleeding in cases of unlocalized subarachnoid hemorrhage. Obviously, the CT scan, if it shows hemorrhage within sulci or intracerebrally, will give a better clue as to the focus of hemorrhage when both an AVM and aneurysm are involved.

The Operation

The patient is positioned so that the malformation is located at the highest point from the heart. It is wise to have the surface of the AVM horizontal in order to facilitate brain relaxation. Large craniotomies are utilized to allow maximum visualization of the normal and abnormal landmarks. The dura is opened carefully so as to precisely divide the numerous adhesions that may occur between components of the malformation and undersurface of the dura. Resection is commenced by a circumscribing incision around the periphery of the malformation as judged by preoperative angiography, especially stereoscopic angiography (Fig. 6). The major arteries supplying the AVM are often located deep in a sulcus adjacent to the margin of the AVM and must be pursued to this point in order to occlude them precisely at the margin of the malformation and not in a position where they may supply normal areas of the brain. Smaller cortical arteries are interrupted on the surface. One major draining vein must be preserved until the final supply to the malformation is interrupted. If this is not done, the malformation will most likely enlarge and explode during the surgical resection. The incision is then deepened around the borders of the malformation, care being taken to identify and secure the deep arterial feeders at an early stage. Veins which drain to the ependymal surface must be carefully handled so as to prevent intraventricular hemorrhage. Finally, the malformation is removed by hinging it on its major venous pedicle and then cauterizing, clipping and dividing this final vessel. If hypotension has been utilized, the blood pressure is brought back to normal levels and the bed of the malformation inspected for bleeding points.

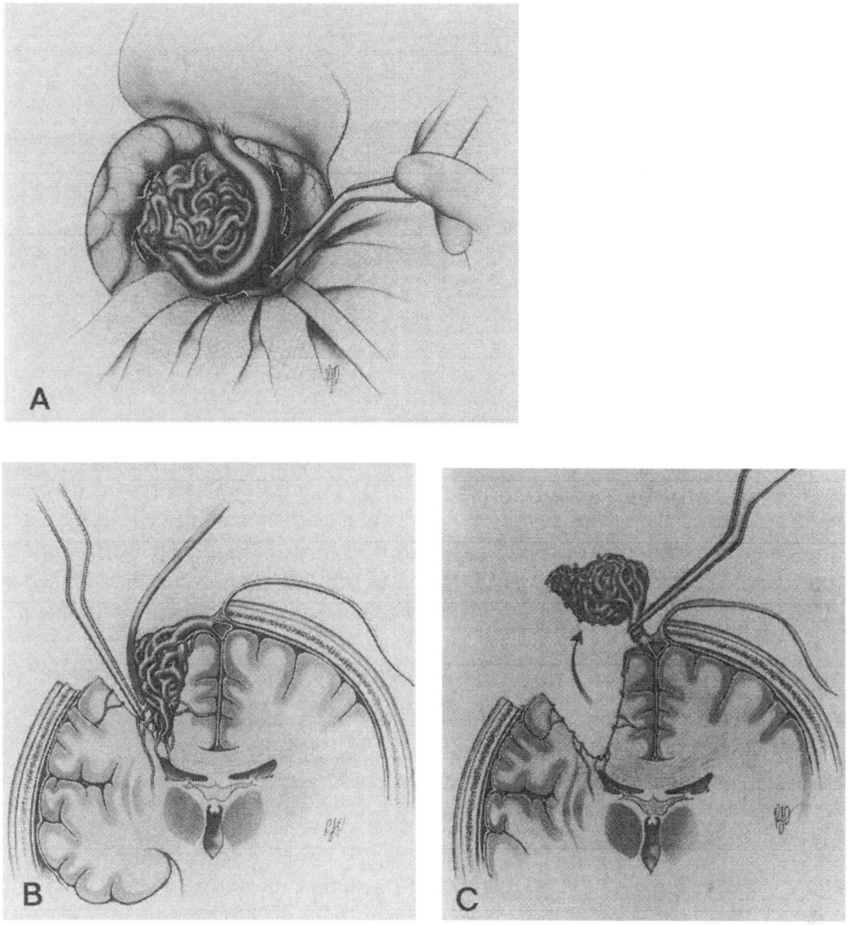

Fig. 6. **A** Drawing showing the planned circumscribing incision around the AVM to secure superficial-cortical arterial supply. The arrows indicate the path of the dissection. **B** Drawing of deep dissection around the margin of an AVM showing control of the deep feeding arteries with preservation of the major draining vein. **C** Drawing showing the final removal of the AVM hinging it upon its large draining vein which is clipped, cauterized and divided

Postoperative Management

Postoperative care includes the control of blood pressure and a CT scan should be obtained if there is any question regarding postoperative hemorrhage. Patients who have had AVMs resected are subject to postoperative seizures and anticonvulsant drugs must be administered. If seizures occur, rapid and high-dose anticonvulsant therapy is indicated.

Of 180 patients with AVMs treated surgically, the mortality rate has been 1.5%. Overall morbidity was about 10% but this includes hemianopsia after the removal of occipital malformations. Significant morbidity, including aphasia, hemiparesis and hemisensory loss is the range of 5%.

Spinal AVM

Summary

Personal surgical experience with approximately 40 arteriovenous malformations of the spinal cord reveals an incidence of approximately 20% of the lesion which were significantly intramedullary (Stein 1979; 1983). We have not found embolization as useful in the preoperative management of AVMs of the spinal cord.

Pathophysiology

Arteriovenous malformations of the spinal cord have been categorized basically into 3 types with one subtype. 3 main types include: (1) A collection of arterialized veins situated on the dorsal surface of the cord. (2) A nidus with arterial to venous fistulae located at a segmental region of the spinal cord. (3) A rare group of malformations which permeate the entire aspect of the spinal cord. The subgroup includes those malformations usually of the nidus type which are located significantly within the spinal cord.

These malformations produce symptomatology by hemorrhage (quite frequent, in intramedullary lesions, rare in the other lesions) by vascular steal and by mass affect. Their course is generally progressive with variable development of severe neurological deficits of a cryptic nature.

The Role of Preoperative Embolization

Since these lesions are low flow and generally treated satisfactorily by surgery, the role of embolization has played diminishing importance. Contrary to cerebral embolization, the backbone of spinal AVM embolization has been the use of liquids which solidify upon reaching the blood stream. These are injected into the malformation by the laborious technique of spinal angiography. It has been found difficult to control the flow of these solidifying substances and prevent their fragmentation. It is for this reason and the fact that most of these malformations are low flow variety that this technique has found usefulness in the preoperative treatment of these malformations as compared with the cerebral malformations.

Surgery

Operations are carried out on the basis of arteriographic, CT and myelographic data. Those malformations in nidus form or involving arterialized veins over the dorsal surface of the cord, must be removed to the point where the surgeon is satisfied that there are no longer arterial to venous shunts. The malformations which are located intramedullary are removed by the standard technique of intramedullary surgery developed in the treatment of intramedullary tumors.

Results

The results in this small series has been most gratifying in those cases of intramedullary AVMs. The juvenile or AVM that permeates the spinal cord is generally considered inoperable and those associated with coils of arterialized veins on the dorsal surface of the cord have had less dramatic results. The nidus form located on the dorsal surface of the cord has responded in a gratifying fashion to surgical obliteration.

References

Amacher AL, Allock JM, Drake CG (1972) Cerebral angiomas: the sequelae of surgical treatment. J Neurosurg 37:571–575

Drake CG (1979) Cerebral arteriovenous malformations: considerations for and experience with surgical treatment in 166 cases. Clin Neurosurg 26:145–208

Garrido E, Stein BM (1978) Removal of an arteriovenous malformation from the basal ganglion. J Neurosurg Psychiatr 41:992–995

Guidetti B, Delitala A (1980) Intracranial arteriovenous malformations. Conservative and surgical treatment. J Neurosurg 53:149–152

Juhasz J (1978) Surgical treatment of arteriovenous angiomas localized in the corpus callosum, basal ganglia and near the brain stem. Acta Neurochir 40:83–101

Kunc Z (1974) Surgery of arteriovenous malformations in the speech and motor sensory regions. J Neurosurg 40:291–303

Luessenhop AJ, Presper JH (1975) Surgical embolization of cerebral arteriovenous malformations through internal carotid and vertebral arteries. Long term results. J Neurosurg 42:443–451

Michelsen WJ (1979) Natural history and pathophysiology of arteriovenous malformations. Clin Neurosurg 26:307–313

Morello G, Borghi GP (1973) Cerebral angiomas. A report of the 154 personal cases and a comparison between the results of surgical excision and conservative management. Acta Neurochir 28: 135–155

Pool JL, Potts DG (1965) Aneurysms and arteriovenous anomalies of the brain. Diagnosis and treatment. New York, Harper and Row 1965, pp 463

Stein BM (1979) Arteriovenous malformations of the brain and spinal cord. Practice of Surgery, Chap 17. Maryland, Harper and Row Publ Inc

Stein BM (1983) Intramedullary spinal cord tumors. Clinical Neurosurgery. Congress of Neurological Surgeons, Chap 37, 30:717–741

Stein BM, Wolpert SM (1977) Surgical and embolic treatment of cerebral arteriovenous malformations. Surg Neurol 7:359–369

Stein BM, Wolpert SM (1980a) Arteriovenous malformations of the brain. I. Current concepts and treatment. Arch Neurol 37:1–5

Stein BM, Wolpert SM (1980b) Arteriovenous malformations of the brain. II. Current concepts and treatment. Arch Neurol 37:69–75

Troupp H (1976) Arteriovenous malformations of the brain: what are the indications for operation? In: Morley TP (ed) Current controversies in Neurosurgery. Saunders, Philadelphia, pp 210–216

Wilson CB, Dominque J (1979) Microsurgical treatment of intracranial vascular malformations. J Neurosurg 51:446–454

Wolpert SM, Stein BM (1975) Catheter embolization of intracranial arteriovenous malformations as an aid to surgical excision. Neuroradiology 10:74–85

Wolpert SM, Stein BM (1979) Factors governing the course of emboli in the therapeutic embolization of cerebral arteriovenous malformations. Radiology 131:125–131

Yasargil MG, Jain KK, Antic J, Laciga R (1976) Arteriovenous malformations of the splenium of the corpus callosum: microsurgical treatment. 5:5–14

Application of the Argon Laser in Neurosurgery

Y. Hosobuchi

An argon ion surgical laser system (Model 770 AMPL, Cooper Medical Corporation, Mountain View, California) has been used in our department since 1981. This laser operates in the blue-green portion of the visible spectrum between 488 and 514.5 nm; approximately 80% of the power output is available at these wavelengths. The laser is coupled to the operating microscope via a fiberoptic micromanipulator delivery system. Transmission through the fiberoptic cables results in a rather homogeneous distribution of energy within the laser beam. The direction and spot size of the beam are controlled, respectively, by a "joy stick" and a focus knob on the micromanipulator. The joy stick can decrease physiological tremor by a factor of 20 : 1. The laser beam spot size can be continuously varied between 0.15 and 1.5 mm. The power output, pulse duration and repetition rate of the laser are controlled from the console.

Because the wavelength of the argon laser is in the visible portion of the spectrum, operating room personnel must wear protective (tinted) eyewear when the laser is in use. An automatic shutter system is incorporated into the micromanipulator-microscopic attachment so that protective eyewear is not necessary for the surgeon or assistant when the surgical microscope is being used. The fact that the beam is visible, however, is advantageous. The CO_2 laser beam is invisible, and a helium-neon laser is used as a visible aiming beam; in that case, the two beams must be critically aligned to assure maximum precision of operation. With the argon laser, this is not necessary, and thus it is intrinsically more precise than the CO_2 laser.

Our clinical experience with the argon laser has been very similar to the experience with the CO_2 laser reported by others. However, each laser has characteristics that dictate its clinical use. The argon laser does not provide high power densities at spot sizes greater than 2 mm in diameter, which are necessary for rapid vaporization of a large, fibrous, or calcified tumor. For this reason, the CO_2 laser is far more efficient for debulking a tumor such as a fibrous avascular meningioma. However, light from the argon laser is selectively absorbed by hemoglobin, which enhances coagulation and vaporization of vascular tumors such as hemangioblastoma. The beam is also transmitted through water and therefore can be used easily for hemostasis while irrigating fluid is applied to the bleeding point. In addition to tumor excision, we have found that argon laser is especially well-suited for making fine incisions in neural tissue (e.g., myelotomy, corpus callosal section) and for making discrete, reproducible lesions in the dorsal root entry zone of the spinal cord without charring.

Listed in Table 1 are some of the pathological conditions for which the argon laser has been used in our department (Powers et al. 1984). These lesions are categorized according to the usefulness of the laser during the surgical procedure.

Table 1. Usefulness of the argon laser in operations for various pathological conditions

Very helpful
Dorsal root entry zone lesion
Hemangioblastoma
Spinal intramedullary tumors
Third ventricular tumor by corpus callosal section
Vascular meningioma
Fourth ventricular ependymoma
Medulloblastoma
Choroid plexus tumor
Cranial nerve schwannomas
Pituitary adenoma
Cerebellar astrocytoma
Helpful
Syringomyelia
Metastatic tumor
Craniopharyngiomas
Fibrous meningioma
Not helpful
Malignant gliomas
Arteriovenous malformations

The argon laser was considered to be very helpful when the surgical result was judged to be better than that obtained with standard microsurgical technique; helpful when the result was equivalent; and not helpful when the result was inferior.

Because technological advances are being made rapidly in medical laser research, it is likely that many more potential applications of lasers will be found. More experimental work and clinical experience will be needed to fully understand the biological effects of these instruments so they can be used for the maximum benefit of our patients.

Reference

1. Powers SK, Edwards MSB, Boggan JE, Pitts LH, Gutin PH, Hosobuchi Y, Adams J, Wilson CB (1984) Use of the argon laser in neurosurgery. J Neurosurg 60:523–530

The Treatment of Brain Tumors by Photoradiation

E. R. Laws Jr., R. E. Wharen Jr., and R. E. Anderson

The concept of photoradiation therapy (PRT) is based upon the ability of certain substances known as photosensitizers to concentrate preferentially in malignant tissue. These photosensitizers then have the capability for the selective destruction of malignant tissue, when activated by light of the appropriate wavelength and intensity in the presence of oxygen. Phototherapy, photodynamic therapy, and photochemotherapy are all terms which have been used so refer to this phenomenon of photoradiation therapy.

The action of a photosensitizer is produced by the absorption of photons of a wavelength sufficient to promote electrons within the sensitizer to an excited triplet state. This excited molecule may then interact either directly with substrates within a cell or indirectly with those substrates through the production of singlet oxygen ($^1O_2$). The various photochemical reactions that are excited by light are subsequently capable of killing cells through multiple interactions with the cell membrane, cytoplasm, nuclear membrane and nucleus.

An ideal photosensitizer should have a number of properties: (1) it must be nontoxic to normal tissues (2) it should be selectively absorbed or retained by all neoplastic or dysplastic cells, (3) it should have some characteristic such as fluorescence that makes it easily detectable, and (4) it must be efficient in killing malignant tissue following the application of light at a wavelength capable of significant tissue penetration.

The search for an ideal photosensitizer is currently being pursued. Although far from ideal, the photosensitizer which has received the most extensive investigation both in the laboratory and clinically is hematoporphyrin derivative (HpD) (Fig. 1).

History of Photoradiation Therapy

The first application of hematoporphyrin derivative (HpD) to the management of brain tumors took place in the 1950's at John Hopkins. At that time, a series of experiments were performed to evaluate HpD fluorescence in the detection of brain tumors at the time of surgery. These experiments demonstrated that detectable levels of HpD would concentrate in brain tumors and that HpD was effectively exluded by an intact blood-brain barrier.

The first studies that suggested that photoactivation of HpD might be cytoxic to brain tumors were performed in 1975. In a tumor model system, cell death was produced by exposure of the tumor containing HpD to light. Interest in this phenomenon was rekindled in Italy, Australia and the United States in 1980 when several

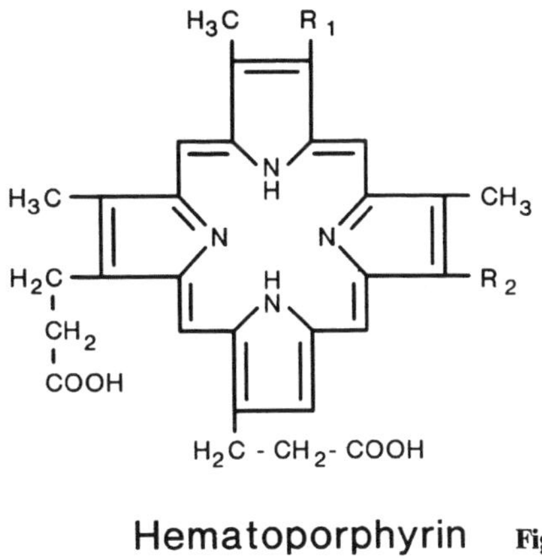

Fig.1

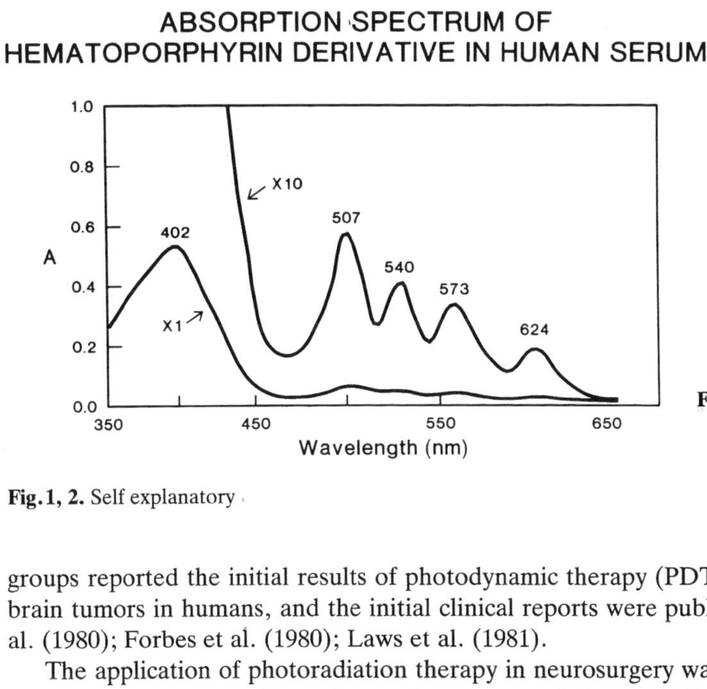

Fig.2

Fig. 1, 2. Self explanatory

groups reported the initial results of photodynamic therapy (PDT) directed toward brain tumors in humans, and the initial clinical reports were published by Perria et al. (1980); Forbes et al. (1980); Laws et al. (1981).

The application of photoradiation therapy in neurosurgery was initially very encouraging. A number of investigators have reported the capability of HpD to selectively kill glioma cells both in vitro and in vivo. More recent reports have described initial attempts in the use of photoradiation therapy for the treatment of malignant brain tumors, and although the results are equivocal, it is evident that despite major differences in the protocols by various investigators, HpD phototherapy is capable of tumor cell destruction in man. However, Rounds and Bonnet both warned that hematoporphyrin is not entirely contained within neoplastic tissue and that some HpD accumulates in brain tissue (Rounds et al. 1982; Bonnet and Berenbaum 1983). This small amount of HpD within normal brain tissue can produce significant morbidity and mortality in experimental animals upon application of a sufficient dose of light.

Efforts to quantitate the uptake of hematoporphyrin within tumors and normal tissue have been performed and Wharen et al. (1983) have quantitated the amount of HpD achieved in human gliomas and normal brain tissue 24–48 h following the systemic administration of 5 mg/kg HpD. The uniformity of brain tumor uptake was recently assessed by Boggan et al. (1984a) who found considerable heterogeneity of uptake in their tumor model, and attributed cell death in such tumors as much to an effect on the vasculature as to direct toxicity to tumor cells.

Current efforts are being directed toward an understanding of the fundamental principles and the scientific application of this modality of photoradiation therapy. This has been stimulated by the desire to develop a treatment which may open new possibilities for therapy of malignant tumors where surgery, radiotherapy, and chemotherapy are inadequate.

Current Status of HpD Photoradiation Therapy

Analysis of the Drug

At the Mayo Clinic in Rochester, Minnesota, HpD was being evaluated for the detection and therapy of bronchogenic cancer under an Investigational New Drug program. Prior studies had demonstrated that administration of the drug to patients at a dosage of 5 mg/kg was safe except for the recognized cutaneous phototoxicity. The Mayo Clinic group then began a phase I study of the feasibility of HpD photodynamic therapy in patients with malignant brain tumors who had failed prior therapy. This series of patients demonstrated the fact that HpD could be administered at a dose of 5 mg/kg without adverse effects. Biopsies of tumor tissue confirmed the fact detectable levels of HpD were present in tumor tissue and that little or no HpD was detectable in adjacent normal brain. Qualitative fluorescent techniques were utilized in reaching these conclusions. It was not possible to confirm with certainty that PDT in these patients had produced a major cytotoxic effect, but some PDT induced necrosis was strongly suggested both by biopsy and subsequent CT scan analysis.

HpD is usually administered clinically (following sterile preparation) as a piggyback infusion at a dose of 3–5 mg/kg over 5–10 min into a freely running intravenous line of D_5 0.2% saline. Thus far, there has been no toxicity associated with the administration of HpD in this manner. There has been some attempt to standardize the preparation of HpD, and indeed the product has recently been marketed as Photofrin I and subsequently Photofrin II (Johnson & Johnson Co., Raritan, NJ).

The absorption spectrum of HpD consists of 5 bands extending from the near-ultraviolet to the red (Fig. 2).

In parallel with the initial clinical investigations, a number of laboratory experiments were performed. These utilized both cell culture models of human and experimental brain tumors and an animal model of malignant ENU-induced brain tumors in rats. These experiments have confirmed HpD photodynamic cytotoxicity for brain tumor cells at practical concentrations of HpD and achievable doses of light. They also provided information relative to the time course and quantitative aspects of HpD concentration within the tumor cells. Several aspects of PDT were elucidated by these studies. First, a dose of 5 mg/kg of HpD seemed adequate. Second, a time delay of 4–24 h resulted in satisfactory concentrations of HpD in the tumor cells and optimal tumor/brain ratios. Third, the exclusion of HpD by an intact blood-brain barrier was again confirmed. Fourth, the superiority of violet light (405 nm) activation to red light activation was confirmed. Fifth, the superiority of red light (633 nm) to violet light penetration of brain and brain tumor was confirmed. More recent experiments have suggested that the postulated cytotoxic mechanism is dependent upon production of singlet oxygen in these experimental brain tumors.

The optimum time to administer photoradiation therapy following drug administration would be when the HpD concentration in tumor versus normal brain is maximum, provided that the absolute amount in normal brain is low enough to be nontoxic. Although Wharen et al. (1983) noted a ratio approximately 2-fold higher at 4 h compared to 24 h after drug administration, Boggan et al. (1984 b), found maximal ratios at 24 h. This ratio will also be dependent upon the HpD preparation, and further development is necessary before the timing of photoradiation therapy after drug administration can be optimized.

The toxicity of HpD in clinical applications has thus far been limited primarily to skin sensitization. Patients must remain out of bright sunlight for approximately 4 weeks following drug administration. McCulloch et al. (1984) have also reported one patient who developed cerebral edema following the administration of 150 joules/cm^2 of red light 48 h after the injection of 5 mg/kg of HpD. As suggested by El-Far and Dimstone (1984) and Dougherty (1984) the use of a more pure preparation of HpD or a different porphyrin such as uroporphyrin-I may limit or eventually eliminate at least the skin toxicity.

Analysis of the Drug-Light Interaction

The action mechanism in photoradiation therapy and in dye-sensitized photo-oxidation reactions for both in vitro and in vivo systems has been an area of avid research from which some understanding of the basic processes involved has developed.

With few exceptions, photosensitized oxidations proceed by way of a triplet sensitizer. The excited triplet state (^3S) of the sensitizer (S) is produced by the absorption of a photon of light with an energy sufficient to raise the sensitizer to an excited singlet state (^1S). Subsequent intersystem crossing results in the transformation of an excited singlet state (^1S) to an excited triplet state (^3S), as the direct excitation of the triplet state (^3S) from the ground state (S) is a forbidden process (Fig. 3).

The excited triplet state (^3S) can then react with biological substrates by 2 major mechanisms: either directly with the substrate (Type I) by electron or hydrogen-ion

Step (1) $S \xrightarrow{h\nu} {}^1S$ Absorption of light

Step (2) ${}^1S \longrightarrow {}^3S$ Intersystem crossing

Fig. 3. Mechanism of the production of an excited triplet state. Absorption of a photon (hv) of light results in the production of an excited singlet state (^1S) which is converted to an excited triplet state through an intersystem crossing

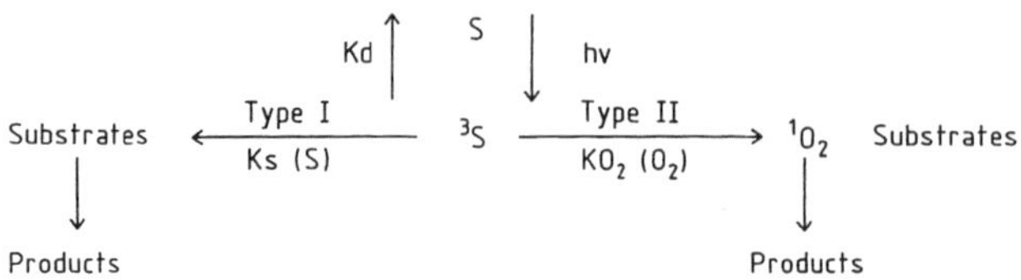

Fig. 4. Mechanism of photooxygenation reactions – an excited triplet sensitizer (^3S) can react with biological substrates either directly (Type I photooxygenation) or indirectly (Type II photooxygenation) through the production of singlet molecular oxygen ($^1O_2$). The efficiency of each path is dependent upon the various concentrations of oxygen (O_2), substrates (S), and triplet sensitizer (^3S), as well as the rate constants for each reaction (KO_2 and Ks) and for the rate of triplet decay (Kd)

State	Symbol	Energy above ground level	Highest Orbital Occupancy
2nd excited state	$^1\Sigma g^+$	37 Kcal	
1st excited state	$^1\Delta g$	22 Kcal	
Ground state	$^3\Sigma g^-$		

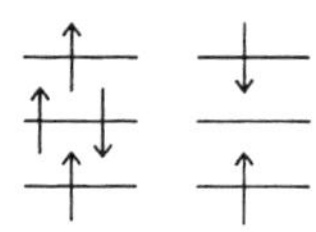

Fig. 5. Electronic configurations of molecular oxygen

transfer *or* indirectly with the substrate (Type II) through the production of singlet molecular oxygen (Fig. 4).

The efficiency of each path is dependent upon the relative concentrations of oxygen (O_2) and substrate (S) available for the sensitizer (^3S) available to react, and the relative rate constants for each reaction (Ko_2 and Ks) and for the rate of triplet decay (Kd). The overall process is limited by the quantum yield of triplet sensitizer formation (O). The rate constants will be dependent upon such factors as the chemical structure of the sensitizer (^3S), the aggregation state of the porphyrin, and the nature of the reaction medium.

Singlet oxygen ($^1O_2$) is a reactive metastable state of oxygen produced by the transfer of energy from an excited triplet sensitizer (^3S) to ground state oxygen (O_2). The electronic structure of an oxygen molecule has 2 highest-energy electrons which are distributed between 2 degenerate or energetically equivalent pi orbitals. If 2 electrons occupy the same orbit, their orbital angular momentum must be in the same direction (a Δ state) and their spins must be opposite resulting in a singlet state. If 2 electrons are in separate orbitals, their angular momenta must be opposite (a Σ state) and their spins can be either opposite ($^1\Sigma$) or parallel ($^3\Sigma$). For oxygen, the lowest energy state is $^3\Sigma$ while the other 2 states ($^1\Delta$ and $^1\Sigma$) are slightly higher in energy, 22 Kcal and 37 Kcal respectively, resulting in 2 energy states of singlet oxygen characterized by different electron distributions within a pair of degenerate molecular orbitals (Fig. 5). The $^1\Delta$ state is long-lived and can survive at least 10^8 collisons while the $^1\Sigma$ state is short-lived and survives no more than 10 collisions under similar conditions. The principal oxygen product of most type II dye-sensitized photo-oxygenation reactions is thus the $^1\Delta$ state of singlet oxygen.

The transfer of energy from the excited triplet state of the sensitizer (^3S) to oxygen proceeds by a process of electronic or resonance energy transfer, which is a diffusion controlled process. Cannistrade et al. found that the $^1O_2$-producing ability of photoexcited porphyrins and hence their photosensitizing efficiency was directly related to the lifetime or the decay rate of the porphyrin triplet state.

The transfer of energy from the excited triplet state (^3S) to oxygen must also involve a consideration of the absorption spectrum of oxygen. As pointed out by Kearns (1969, 1971), for transfer of energy to occur, there must be a close overlap of the oxygen absorption peak with the energy spectrum of the donor or excited triplet state. O_2 has multiple absorption peaks within the visible spectrum between 400–800 nm including one peak at 634 nm. This peak is coincidentally very close to the normal wavelength of a He-Ne laser (632 nm) which has been used for photoradiation therapy. Additional peaks occur also in the near-infrared region between 1–2 μ, and Evans (1969) has reported the efficient production of $^1O_2$ and subsequent photooxidation reactions occurring at near-infrared wavelengths.

Spikes demonstrated in 1975 that porphyrin-sensitized photooxidations involved the lowest excited triplet state of the dye as the reactive intermediate. The lowest excited triplet state for hematoporphyrin (^{3}Hp) is formed with a quantum yield of 0.8 or greater.

Dougherty et al. proposed in 1976 that singlet oxygen ($^1O_2$) was the cytotoxic agent responsible for the in vitro inactivation of TA-3 mouse mammary carcinoma cells exposed to HpD photoradiation using red light therapy. $^1O_2$ was produced by the transfer of energy from an excited triplet state of hematoporphyrin to oxygen with a quantum yield for $^1O_2$ of 0.16 within the TA-3 cells. Further work has demonstrated that although most porphyrin-sensitized reactions occur via a Type II mechanism involving $^1O_2$, other Type I mechanisms involving electron transfer may be important. Grossweiner et al. (1982) reported that lysis of phosphatidylcholine liposomes was produced either via a Type I or Type II mechanism depending upon the concentration of Hp and HpD, and suggested that Type II processes mediated reactions involving nonaggregated Hp while Type I processes mediated reactions involving aggregated Hp and reactions within hypoxic tumor tissue. Type I mechanisms became more prominent at higher concentrations of HpD where there is a greater tendency of HpD to form aggregates. Since the active component of HpD is considered to be aggregated hematoporphyrin, both types of mechanisms may be important for in vivo reactions. Indeed, Moan has documented that HpD (Photofrin I and II) which consists of porphyrin aggregates produces $^1O_2$ with significantly smaller yields than Hp. Thus, although the use of $^1O_2$ quenchers such as β-carotene, ascorbic acid, and N_3- plus the enhancing effect of D_2O on the production of $^1O_2$ have clearly demonstrated the participation of $^1O_2$ in reactions in vivo, no photodynamic action so far investigated in vivo is solely explained by the $^1O_2$ mechanism.

The mechanisms responsible for the loss of cell viability in photoradiation therapy are difficult to characterize because of the multiplicity of damaging reactions that occur. Photoradiation of porphyrin-loaded cells results in inhibition of membrane transport functions and membrane damage; effects which may occur from photodynamic cross-linking of membrane proteins. Photodynamic damage to DNA and to lysosomes also occurs.

Proteins, nucleic acids, unsaturated lipids NADH, NADPH, hyaluronic acid and other biomolecules are photo-oxidized with porphyrins as sensitizers. The predominate susceptible sites on proteins are the unprotonated thiol group of cysteine, the unprotonated imidazole ring of histidine, the thio-ether group of methionine, the indole ring of tryptophan, and the phenolate anion of tyrosine. Photo-oxidation of these sites on proteins results in the loss of enzymatic and hormonal activity, loss of toxic properties of snake venoms and bacterial toxins, loss of antigenic properties, and loss of antibody reactivity. Unsaturated lipids and cholesterol are converted to hydroperoxides. Nucleic acids are photo-oxidized by porphyrins predominately at guanine residues, and both single and double strand breaks can be produced in DNA.

Inactivation of cells by photoradiation can be classified into three major modes of action. First, the porphyrin remains either outside the cell or within the cell membrane. In this case, the cell membrane would be expected to be the major site of photodamage. Membrane damage would result from the photo-oxidation of membrane lipids, structural proteins, and enzymes with resultant inhibition of transport processes, alterations of receptors, changes in permeability, or cross-linking of proteins in the membrane. Second, the porphyrin penetrates into the cytoplasm resulting in photodamage to mitochondria, lysosomes, ribosomes, and proteins. Cell death would then occur from uncoupling or inhibition of oxidative phosphorylation, leakage of hydrolases from lysomes into the cytoplasm, and inhibition of microsomal activity. Third, the porphyrin penetrates the nucleus to sensitize the nucleic acids and chromosomes with resulting chromosomal breaks. As yet, no chromosomal breaks have been demonstrated in mammalian cells. Currently, the modes of action of porphyrin photosensitized reactions are considered to be multifactorial involving predominately the cell membrane and cytoplasm.

Optimization of the parameters of photoradiation therapy involves not only considerations of the uptake, distribution, and action mechanism of the dye but also considerations of the wavelength, quantity, and energy density of light necessary to achieve cell kill. The action spectrum for porphyrin sensitized cytotoxicity corresponds closely to the absorption spectrum of the porphyrin. Kinsey et al. (1981) found that the cytotoxic action of HpD was directly proportional to the number of light quanta absorbed by HpD in the cell. For thin layers of cells, the Soret band at 405 nm had 12–30 times the cytotoxicity of red light.

Anderson et al. (1984) investigated the effect of optical spectrum, power density, HpD concentration, and HpD preparation on the HpD tumor cell killing efficiency of MEWO cells in culture. Using a trypan blue exclusion assay, cell survival curves were obtained following irradiation with violet (405 nm), red (630 nm), and white light (340–680 nm) at energies of 0–320 joules and at power densities of 20–160 mW/cm^2 for cells exposed for 6 hours to HpD concentrations of 0–25 μg/ml. All studies were performed for both Photofrin I and for Mayo Clinic preparations of HpD (Fig. 6).

The survival curves demonstrated a consistant pattern with an increasing slope and decreasing shoulder to the curve at increasing concentrations of HpD. Identical curves were obtained for both Mayo Clinic HpD and Photofrin I. The ratio of the killing efficiency of violet light to red light was approximately 20:1. For red light, a 2-fold increase in cellular killing efficiency was observed at higher power densities of 160 mW/cm^2 compared to a lower power density of 40 mW/cm^2. 110 joules of red light at a power density of 40 mW/cm^2 produced a 50% cell kill compared to only 55 joules required at a power density of 160 mW/cm^2, at an HpD concentration of 25 μg/ml. This phenomenon of an increased killing efficiency of red light at lower power

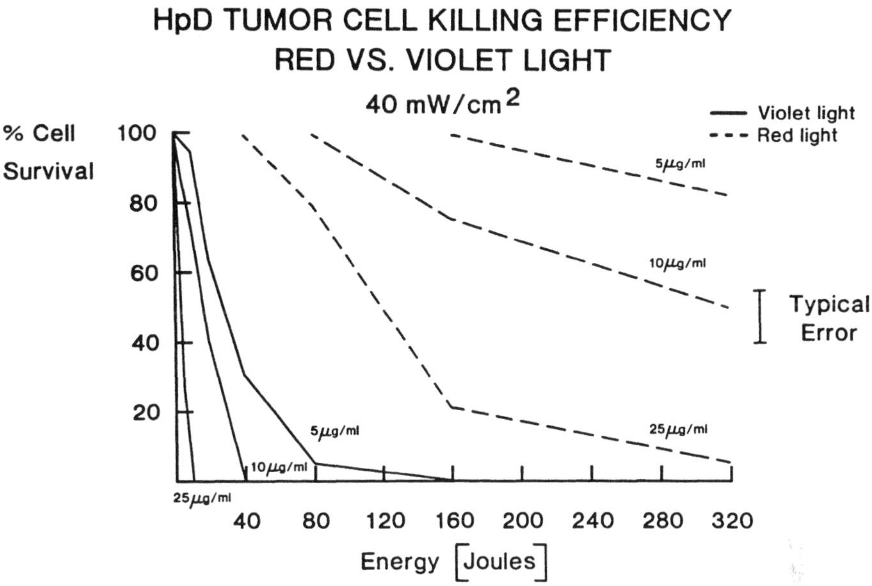

Fig. 6. Self explanatory

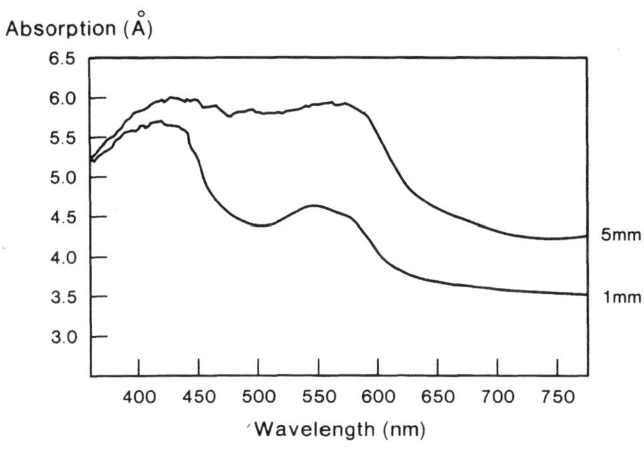

Fig. 7. Absorption spectrum of light of various wavelengths in cat brain at depths of 1 and 5 mm

densities has also been observed by Dougherty (1984) who attributed it to a partial repair process occurring during exposure. Gomer et al. (1984), however, found that the cytotoxicity of CHO cells in cultures was independent of the dose rate of red light at lower power densities of 0.5 to 60 mW/cm^2. Using 2.5 nm increments of red light from 620–640 nm, the most efficient wavelengths of red light for cell killing were 627.5 to 632.5 nm.

For violet light, however, a marked decrease in cellular killing efficiency is found at higher power densities. This decrease in cellular killing efficiency at higher power densities varied greatly with concentration, being more pronounced at lower concentrations of HpD. This process was demonstrated to be irreversible as cells irradiated at a high power density of white light followed by irradiation at low power densities of violet light still demonstrated a markedly decreased cellular killing efficiency compared to cells irradiated only at a low power densities of violet light. This phenomenon suggested an irreversible photodecompositon of HpD at higher power densities of violet light.

Not only the wavelength and power density but also the type of light appears to effect the cellular killing efficiency of HpD-photoradiation therapy. Cowled et al. (1984) observed no difference in the HpD cell killing efficiency of a continuous wave argon pumped dye laser using Rhodamine B with a wavelength of 625–635 nm compared to a pulsed-wave Gold vapor laser at a pulse frequency of 10–14 KHz having a wavelength of 627.8 nm. Wharen et al. (unpublished data) found a markedly decreased cellular killing efficiency for pulsed red light (625–645 nm) produced from a tunable flash-pumped dye laser at a repetition rate of 1–6 s compared to continuous wave red light (625–635 nm) from a filtered xenon arc lamp. In addition, Andreoni et al. (1982) have reported that pulsed light from a nitrogen laser with a wavelength of 332 nm and a repetition rate of 30 Hz had a HpD cellular killing efficiency greater than a continuous light from an argon ion laser with a wavelength of 334 nm. They attributed this to a mechanism of two photon absorption and production of cytotoxic radicals of HpD. It appears that not only the wavelength of light but also the form of the light (pulsed vs. continuous) and the repetition rate, pulse width, and peak pulse power are all important variables in need of further study.

After all the drug and light parameters have been maximized in HpD photoradiation therapy, the limiting factor in its clinical application may remain the penetration of light through brain and tumor tissue. Light penetration into tissue is determined by the optical characteristics of the tissue, the wavelength of the light, and the concentration of the photosensitizer that has been used. Photons are either absorbed or scattered and the ultimate penetration of light is both wavelength and tissue dependent in an exponential manner.

The relative penetration of light through in vivo cat brain as a function of wavelength has been measured and demonstrates the significantly greater depth of penetration of red light (630 nm) compared to violet light (405 nm) (Fig. 7). Dougherty (1984), however, has recently stated that the useful penetration of visible light in adult brain at 630 nm is on the order of 1–1.5 mm. If that is the case, then the depth of penetration of light at 630 nm represents a significant limiting factor for the use of photoradiation therapy in brain and brain tumors. It is known, however, that the penetration of light through tissue continues to improve by several orders of magnitude as the wavelength increases from approximately 600 nm–1.1 m. Thus, the possibility exists that photoradiation therapy at these wavelengths might provide a more effective depth of tissue penetration.

Clinical Studies

Clinical experience (Table 1) has expanded slowly, but steadily. The majority of patients have had recurrent malignant tumors and had failed prior attempts at therapy.

Table 1. Brain tumor patients treated by HpD-photodynamic therapy

Tumor	Location	HpD	Light	Result
Malignant astrocytoma	Right frontal	5 mg/kg IV, 48 h	Laser – 630 nm	Alive, 6 months
Malignant astrocytoma	Left frontal	5 mg/kg IV, 72 h	Laser – 630 nm	Dead, 5 months
Malignant astrocytoma, cystic	Left temporal	5 mg/kg IV, 48 h	Laser – 630 nm	Dead, 37 months
Malignant astrocytoma	Right frontal	5 mg/kg IV, 48 h	Laser – 630 nm	Dead, 4 months
Malignant astrocytoma	Left temporal	5 mg/kg IV, 36 h	Post resection xenon lamp – 405 + 630 nm	Dead, 7 months
Malignant astrocytoma	Left parietal	5 mg/kg IV, 24 h	Post resection xenon lamp – white	Dead, 19 months infection
Malignant astrocytoma	Right frontal	5 mg/kg IV, 24 h	Post resection xenon arc – white	Dead, 26 months
Malignant astrocytoma	Right thalamus	5 mg/kg IV, 24 h	Post resection xenon arc – white	Alive, 33 months
Malignant astrocytoma	Right thalamus	5 mg/kg IV, 6 h	Post resection xenon arc – 630 nm	No FU
Malignant astrocytoma	Left frontal	5 mg/kg IV, 6 h	Post resection xenon arc – 630 nm diffusion medium	Dead, 2 months
Malignant astrocytoma	Left frontal	5 mg/kg IV, 8 h	Post resection xenon arc – 630 nm diffusion medium	Dead, 2 months
Malignant astrocytoma	Left frontal	5 mg/kg IV, 8 h	Post resection xenon arc – 630 nm diffusion medium	Alive with recurrence, 13 months
Malignant oligodendroglioma	Right frontal	5 mg/kg IV, 48 h	Post resection xenon arc – 405 + 630 nm	Dead, 3 months

Table 1 (continued)

Tumor	Location	HpD	Light	Result
Malignant small cell neoplasm (probably metastatic)	Right frontal	5 mg/kg IV, 48 h	Laser – 630 nm	Dead, 1 month infection
Malignant small cell neoplasm (probably metastatic)	Right frontal	5 mg/kg IV, 48 h	Laser – 630 nm	Alive, 6 months
Medulloblastoma	Cerebellum	5 mg/kg IV, 6 h	Post resection xenon arc – white	Alive, 31 months
Medulloblastoma	Cerebellum	5 mg/kg IV, 6 h	Post resection xenon arc – 630 nm diffusion medium	Alive, 13 months
Ependymoma	Cerebellum	5 mg/kg IV, 4 h	Post resection xenon arc – 405 nm	Dead postop. DIC
Ependymoma	Cerebellum	5 mg/kg IV + topical, 6 h	Post resection xenon arc – 630 nm diffusion medium	Alive, 8 months
Metastatic carcinoma (pulmonary)	Left parietal	5 mg/kg IV, 6 h	Post resection xenon arc – 630 nm diffusion medium	Alive with recurrence, 11 months
Metastatic melanoma	Left frontal	5 mg/kg IV, 10 h	Post resection xenon arc – 620 nm diffusion medium	Alive, 9 months
Rhabdomyosarcoma	Left orbit	5 mg/kg IV, 24–72 h	Xenon arc – 405 + 630 nm	Dead, 2 months
Craniopharyngioma	Sella turcica	Topical 5 mg/cc, 15 min	Xenon arc – 630 nm	Dead, pulmonary embolus, 4 months

The mean survival for those patients who died after surgery was 11.6 months. 7 patients are alive and well, and 2 are alive with symptomatic recurrences.

At present, a number of modalities of PDT for malignant brain tumors are suggested. For inoperable deep tumors, the stereotactic implantation of one or more quartz fibers to provide Argon-dye laser photoradiation (500–1000 J) has been utilized, in conjunction with prior intravenous administration of HpD. For recurrent tumors which can be resected, PDT of the tumor bed has been utilized after intravenous, and in some cases additional topical administration of HpD. The light delivery system has consisted of a filtered high-intensity xenon-arc lamp and a fiberoptic cable with a Lucite tip. The latter is inserted into a diffusion medium (0.1% Liposol in Saline) which fills the tumor bed, and also may be used to cool the operative field. A dose of 150–200 J is employed. A third mechanism of PDT is employed for cystic or cavitary lesions. After intravenous and/or topical administration of HpD, the cyst or cavity is filled with a diffusion medium and illuminated either with the laser-quartz fiber system or the high-intensity xenon-arc lamp-fiberoptic system (Fig. 8, 9).

The technical aspects of these clinical trials in brain tumor patients have been quite satisfactory. There have been no adverse effects related to either intravenous or topical administration of HpD.

The light delivery systems have functioned well, but the importance of temperature monitoring should be emphasized. Power densities greater than 200 mW of red light through a 0.6 mm quartz fiber will produce significant heating of tissue. There is less significant heating when light is delivered through a large diameter (>5.0 mm) fiberoptic system. Because hyperthermia has its own cytotoxic effects, and because heat interferes with the photodynamic effect it is essential to monitor this parameter and to avoid any significant thermal effects while delivering photoradiation (Fig. 10).

If tissue heating is avoided, we have not recognized any significant degree of post-therapy cerebral edema in any of these patients, all of whom had CT scans within 48 h of treatment. 2 patients had new neurologic signs related either to surgery or PDT, but they were transient in both. All patients were managed with pre- and postoperative corticosteroid therapy. 2 patients developed postoperative wound infections which were not surprising in light of the extensive prior therapy both had received. Both responded to antibiotic therapy. One patient died postoperatively of disseminated intravascular coagulopathy (DIC) and one late death occurred at 3 months from a pulmonary embolus. 2 patients developed symptoms and signs of cutaneous photosensitivity as a result of disregarding advice to protect themselves from direct sunlight.

As mentioned earlier, the analysis of this series of brain tumor patients treated by PDT with HpD does not yet permit any conclusions as to the effectiveness of the method. The results, however, are encouraging for several reasons. The theoretical basis of PDT for brain tumors still appears to be sound. The administration of effective amounts of HpD is relatively well tolerated by these patients. The light delivery systems are practical in use. At least some of these patients appear to have derived some benefit from PDT.

As basic knowledge with regard to PDT and malignant brain tumors increases and further laboratory work in cell culture and animal model systems progresses, it should be possible to improve the efficiency and the efficacy of PDT. Clinical indications may expand as well, to include the treatment of non-malignant but invasive

Fig. 8

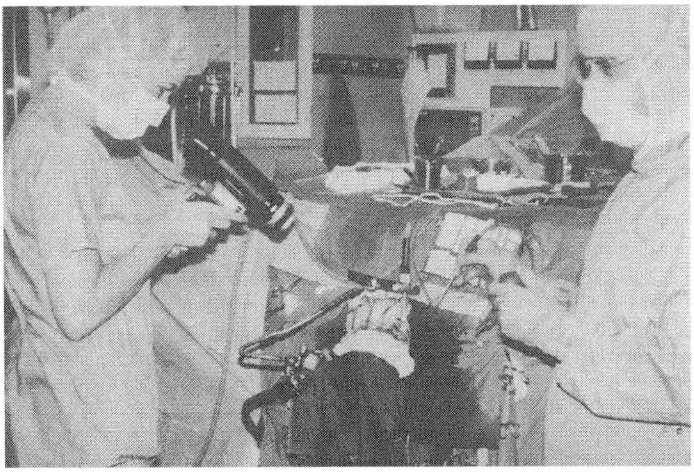

Fig. 9

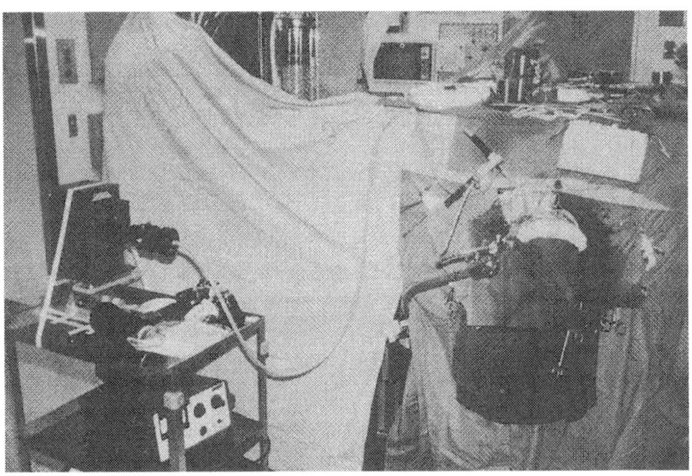

Fig. 10

Fig. 8. High-intensity xenon arc lamp and fiberoptic delivery system
Fig. 9. Detection of fluorescence from brain and tumor with low-light multichannel plate amplifier
Fig. 10. Photoradiation arrangement in operating room. Post resection mode

brain tumors such as meningiomas, pituitary adenoma and craniopharyngioma, and the treatment of some forms of brain abscess or parasitic infestation. Other prospects for the future include the following: new light-drug combinations of higher efficiency, less toxicity, and deeper penetration of tumor tissue; new light delivery systems, such as multiple fiber lasers, for more complete photoradiation of tumor tissue; the development of methods (systemic or topical) to improve dye uptake by tumor cells; the use of metabolic enhancers of the cytotoxic effect in tumor tissue and quenchers of the photodynamic effect in normal tissue; and combination of PDT with other methods (stereotatic CO_2 laser resection, interstitial and conventional radiation therapy, chemotherapy, hyperthermia, immunotherapy) in an effort to achieve ultimate control of these devastating tumors.

Acknowledgment. The authors are grateful to Mrs. Constance B. Hoeft for her expert assistance in the preparation of the manuscript. Portions of this manuscript will appear in Book Chapters from the Clayton Symposium and the meeting on Porphyrins in Alghero, Sardinia in May 1985.

References

Anderson RE, Wharen RE Jr, Jones CA, Laws ER Jr (1984) Parameters of hematoporphyrin derivative tumor cell killing efficiency: Decomposition of hematoporphyrin derivative at high power densities. In: Porphyrin localization and treatment of treatment of tumors. Alan R Liss, Inc, New York, pp 483–500

Andreoni A, Cubeddu R, De Silvestri S, et al (1982) Two step laser activation of hematoporphyrin derivative. Chemical Physics Letters 88:37–39

Boggan JE, Berns M, Edwards M (1984a) Uptake, distribution, and retention of hematoporphyrin derivative in metastatic and intrinsic rat tumor models. In: Porphyrin localization and treatment of tumors. Alan R Liss, Inc, New York

Boggan JE, Walter R, Edwards M, et al (1984b) Distribution of hematoporphyrin derivative in the rat 9L gliosarcoma brain tumor analyzed by digital video fluorescence microscopy. J Neurosurg 61:1113–1119

Bonnett R, Berenbaum MC (1983) HpD – A study of its components and their properties. In: Kessel D, Dougherty TJ (eds) Porphyrin photosensitization. Plenum Publishing Corp, New York, pp 241–260

Cowled PA, Grace JR, Forbes IJ (1984) Comparison of the efficiency of pulsed and continuous wave red laser light in induction of phototoxicity by hematoporphyrin derivative. Photochem Photobiol 39:115–117

Dougherty TJ (1984) Photodynamic therapy (PDT) of malignant tumors. CRC Critical Reviews in Oncology/Hematology 2:83–116

Dougherty TJ, Gomer CJ, Weishaupt KR (1976) Energetics and efficiency of photoinactivation of murine tumor cells containing hematoporphyrin. Cancer Res 36:2330–2333

El-Far MA, Dimstone NR (1984) Superiority of uroporphyrin I over other porphyrins in selective tumor localization. In: Porphyrin localization and treatment of tumors. Alan R Liss, Inc, New York

Evans DF (1969) Oxidation by photochemically produced singlet states of oxygen. J Chem Soc Sect D 8:367–368

Forbes IJ, Cowled PA, Leong AS, et al (1980) Phototherapy of human tumors using hematoporphyrin derivative. Med J Aust 2:489–493

Gomer CJ, Doiron DR, Dunn S, et al (1984) Examination of action spectrum, dose rate, and mutagenic properties of hematoporphyrin derivative photoradiation therapy. In: Porphyrin localization and treatment of tumors. Alan R Liss, Inc, New York

Granelli SG, Diamond I, McDonagh AF, et al (1975) Photochemotherapy of glioma cells by visible light and hematoporphyrin. Cancer Res 35:2567–2570

Grossweiner CI, Patel AS, Grossweiner JB (1982) Type I and type II mechanisms in the photosensitized lysis of phosphatidylcholine liposomes by hematoporphyrin. Photochem Photobiol 36: 159–167

Kearns DR (1969) Selection rules for singlet oxygen reactions, concerted addition reactions. J Am Chem Soc 91: 6554–6556

Kearns DR (1971) Physical and chemical properties of singlet molecular oxygen. Chemical Reviews 71: 395–427

Kinsey JH, Cortese DA, Moses HL, Ryan RJ, Branum EL (1981) Photodynamic effect of hematoporphyrin derivative as a function of optical spectrum and incident energy density. Cancer Res 41: 5020–5026

Laws ER Jr, Cortese DA, Kinsey JH, et al (1981) Photoradiation therapy in the treatment of malignant brain tumors: A phase I (feasibility) study. Neurosurgery 9: 672–678

McCulloch GA, Forbes IJ, Lee See K, Cowled PA, Jacka FJ and Ward AD (1984) Phototherapy in malignant brain tumors. In: Doiron DR, Gomer CJ (eds) Porphyrin localization and treatment in tumors. Alan R Liss, Inc, New York, pp 709–718

Perria C (1981) Photodynamic therapy of human gliomas by hematoporphyrin and He-Ne laser. IRCS Med Sci (Cancer) 9: 57–58

Perria C, Capuzzo T, Cavagnaro G (1980) First attempts at the photodynamic treatment of human gliomas. J Neurosurg Sci 24: 119–129

Rasmussen-Taxdal DS, Ward GE, Figge FHJ (1955) Fluorescence of human lymphatic and cancer tissue following high doses of intravenous hematoporphyrin. Cancer 8: 78–81

Rounds DE, Jacques S, Shelden CH, Shaller CA, Olson RS (1982) Development of a protocol for photoradiation therapy of malignant brain tumors – Part 1. Photosensitization of normal brain tissue with hematoporphyrin derivative. Neurosurgery 11: 500–505

Wise BL, Taxdal DR (1987) Studies of the blood brain barrier utilizing hematoporphyrin: Short communication. Brain Res 4: 387–389

Wharen RE Jr, Anderson RE, Laws ER Jr (1983) Quantitation of hematoporphyrin derivative in human gliomas, experimental central nervous system tumors, and normal tissues. Neurosurgery 12: 446–450

Photodynamic Therapy of Malignant Brain Tumors: Preliminary Experiences with Four Patients

S. Mingrino, G. Jori, and P. Zampieri

The ability of hematoporphyrin (Hp) and some chemical derivatives (HpD) to be accumulated in greater amounts and retained for longer periods of time by neoplastic than by normal tissues is the basis of a new phototherapeutic possibility for tumors (Dougherty 1984) called photodynamic therapy (PDT). The technique has been successfully applied to both superficial (Dahlman et al. 1983) and deep-seated tumors (Dougherty et al. 1981); in the latter case, the light emitted from the laser source is focused into an optical fiber, which is then inserted into the tumor mass. The photoactivation of tumor-localized Hp or HpD is most frequently obtained by irradiation with red light (600–630 nm; Dougherty 1984): these wavelengths have the dual property of not being absorbed by common cell constituents and of having a significant penetrating power into biological tissues. As a consequence, the Hp-promoted photodamage is restricted to the level of the neoplastic areas, which are illuminated in a uniform way for diameters of up to 1 cm (Dougherty et al. 1981).

Recent preliminary reports (Laws et al. 1981) have described the possibility of treatment of malignant brain tumors by PDT. In general, porphyrins cannot cross the intact blood-brain barrier (Wise and Taxdal 1967); however, alteration of the barrier allows ready penetration of the drug, as shown by the observed uptake of Hp by experimental and human gliomas (Perria et al. 1983; Wharen et al. 1983; Boggan et al. 1984) or hypophyseal adenomas (Perria et al. 1985). Apparently, one factor limiting the clinical utilization of PDT in the field of brain tumors is the lack of information on the pharmacokinetic behavior of Hp in humans, in particular as regards the differential accumulation and release of Hp by normal and neoplastic brain tissues.

The presence of Hp or other porphyrins in tumors in vivo is usually investigated by means of fluorescent spectroscopy (Cortese et al. 1979). This technique relies on the intense light emission in the 630–700 nm interval, which is typical of photoexcited porphyrins. However, in spite of recent advances in the sensitivity of the technique following the introduction of digital video fluorescent microscopy (Boggan et al. 1984), this method of analysis at present gives only qualitative information. Actually, the fluorescent yield of porphyrins has been shown (Andreoni et al. 1983) to be controlled by a large variety of parameters, including the aggregation state of the drug and the nature of the microenvironment. The detailed features of these parameters in vivo cannot be readily predicted or evaluated. In consequence, the fluorescent intensity cannot be linearly correlated with the concentration of Hp in a given tissue; moreover, false negatives have been reported (Lipson et al. 1967), due to the presence in neoplastic tissues of porphyrin molecules whose fluorescent emission had been quenched.

We have devised an approach to this problem which should both provide quantitative estimates of the Hp concentration in neoplastic as compared with normal, brain tissues, and also indicate the possibility of a synergistic treatment of brain tumors by a combination of surgery and PDT. Selected patients, who were affected by malignant brain tumors and had to undergo tumor resection, were intravenously injected, after informed consent, with a dose of 5 mg Hp per kg body weight about 24 h before surgery. Samples of the neoplastic and perineoplastic tissues removed were homogenized, and the Hp content was estimated after extraction of the porphyrin by incubation of the homogenate with an aqueous dispersion of sodium dodecylsulphate (SDS) micelles. In this way, the porphyrin is incorporated in a quantifiable way as a monomer into the hydrophobic micellar core (Andreoni et al. 1983). The homogeneity of the physical state and microenvironment of Hp thus obtained eliminates possible artifacts in the spectrophotofluorimetric determination of the porphyrin. As a consequence, the amount of Hp extracted from the tissue can be estimated in quantitative terms by interpolation of the experimentally obtained fluorescent intensity with a suitable calibration curve (Jori et al. 1983).

So far, four patients have been studied. From the data accumulated, the following indications emerge:

1. All tumors examined (three gliomas of different grades and one metastasis) took up the injected Hp in concentrations above 3 mg porphyrin per gram of tissue: this suggests that the affinity of Hp for brain tumors is independent of the tumor type.
2. Perineoplastic tissues gave Hp recoveries at 24 h below 0.3 mg porphyrin per gram of tissue; thus it appears that this time interval is sufficient to achieve a complete clearance of Hp from nonneoplastic areas, yielding a high ratio of Hp concentration between the tumor and normal tissues.
3. The concentration of endogenous porphyrins in brain tissues was found to be less than 0.1 mg of drug per gram of tissue (coincident results were obtained from five tumors removed from patients who had not been previously injected with Hp); therefore, the presence of endogenous porphyrins is far too low to interfere with the fluorescent determination of injected Hp.
4. The treatment is safe and well tolerated by the patients, and no untoward effects were observed. These observations encouraged us to apply PDT for sterilization of the surgical cavity after removal of the tumor.

References

Andreoni A, Cubeddu R, De Silvestri S, Jori G, Laporta P, Reddi E (1983) Time resolved fluorescence studies of hematoporphyrin in different solvent systems. Z Naturforsch 38c: 83–89

Boggan JE, Walter R, Edwards MSB, Borcich JK, Davis RL, Koonce M, Berns NW (1984) Distribution of hematoporphyrin derivative in the rat 9L gliosarcoma brain tumor analyzed by digital video fluorescence microscopy. J Neurosurg 61: 1113–1119

Cortese DA, Kinsey JH, Woolner LB, Payne WS, Sanderson DR, Fontana RS (1979) Clinical application of a new endoscope technique for detection of in situ bronchial carcinoma. Mayo Clin Proc 54: 635–642

Dahlman A, Wile AG, Burns RG, Mason GR, Johnson FM, Berns MW (1983) Laser photoradiation therapy of cancer. Cancer Res 43: 430–434

Dougherty TJ (1984) Urology 23:61–64
Dougherty TJ, Thoma RE, Boyle DG, Weishaupt KR (1981) Interstitial photoradiation therapy for primary solid tumors in pet cats and dogs. Cancer Res 41:61–64
Jori G, Tomio L, Reddi E, Rossi E, Corti L, Zorat PL, Calzavara F (1983) Preferential delivery of liposoma-incorporated porphyrins to neoplastic cells in tumour-bearing rats. Br J Cancer 48: 307–309
Laws ER, Cortese DA, Kinsey JH, Eagan RT, Anderson RE (1981) Photoradiation therapy in the treatment of malignant brain tumors: A phase I (feasibility) study. Neurosurgery 9:672–678
Lipson RL, Baldes EJ, Gray MJ (1967) Cancer 20:2255–2257
Perria C, Delitala G, Francaviglia N, Altomonte M (1983) The uptake of hematoporphyrin derivative by cells of human gliomas: Determination by fluorescence microscopy. IRCS Medical Science 11:46–47
Perria C, Sechi CS, Fresu P, Reddi E, Jori G (1985) Med Biol Environ 13:89–93
Wharen RE, Anderson RF, Laws ER (1983) Quantitation of hematoporphyrin derivative in human gliomas, experimental central nervous system tumors, and normal tissues. Neurosurgery 12: 446–450
Wise BL, Taxdal DR (1967) Studies of the blood-brain barrier utilizing hematoporphyrin. Brain Res 4:387–389

Intraoperative Monitoring with Evoked Potentials

E. R. John, L. S. Prichep, and J. Ransohoff

Neurosurgeons have been monitoring operations with evoked potentials (EPs) for a number of years. Although there is general agreement that such monitoring is potentially very useful, the time required to make a measurement in the intraoperative situation is too long and the variability too great for most practical purposes.

The purpose of this paper is to demonstrate that a new technique which we have developed overcomes these limitations and to summarize our experiences using this method to monitor over 400 surgical procedures of many different types.

An EP is a small transient voltage oscillation time locked to the presentation of a sensory stimulus and imbedded in much larger electrophysiological activity as well as environmental electrical noise. The electrical background varies randomly in time, while the EP does not. For this reason, the background noise can be reduced by averaging many samples of responses to stimulation, unmasking the EP waveshape. As you can see in this example (Fig. 1), averaging reduces masking activity propor-

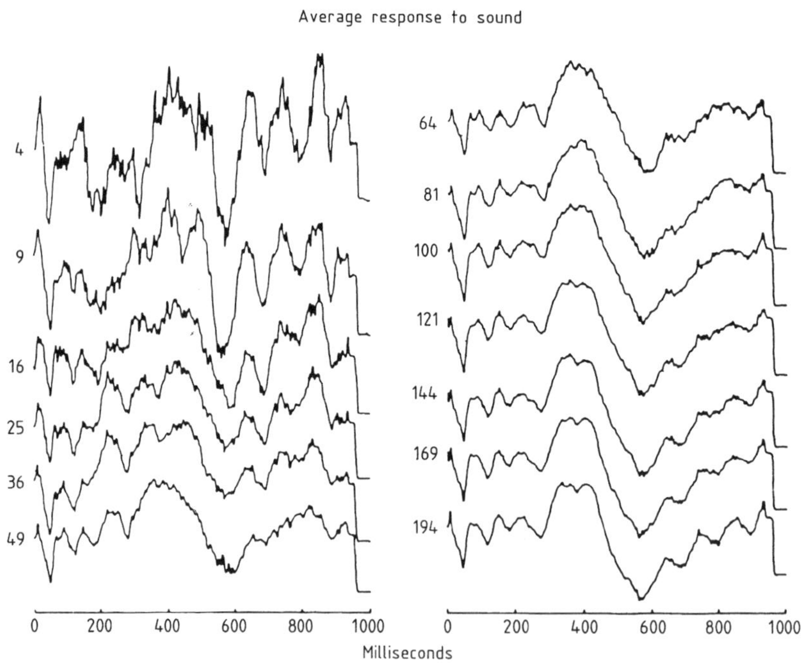

Fig. 1

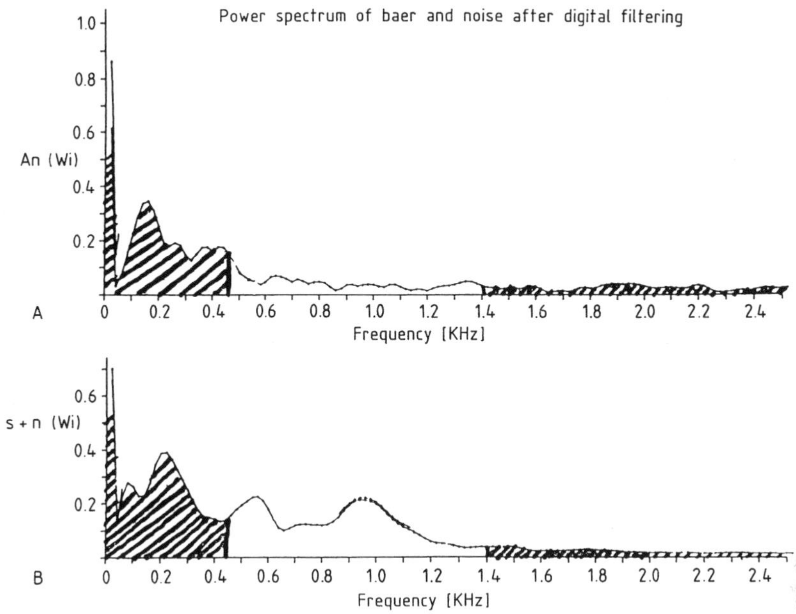

Fig. 2

tional to the square root of the sample size. A large number of samples must be averaged before the emerging EP waveshape stabilizes, but residual noise always distorts the waveshape to some extent. Typically, one to two hundred cortical responses and one to two thousand brainstem or "far field" responses must be averaged to extract the EP waveshape.

The method which we have developed is called optimum digital filtering and has been described by us previously. This method selectively removes most of the masking noise from the evoked response prior to averaging, which accomplishes further reduction of any residual noise. The principle of this method is illustrated in Fig. 2. On top is shown the power spectrum of the background noise recorded from a patient in the absence of stimulation. On the bottom is shown the power spectrum of the noise plus the EP, recorded during stimulation. From these plots of power versus frequency, it is clear that most of the power related to the EP lies only in a part of the frequency range. The shaded regions above and below this range contain as much noise power as signal power. Once the frequency domain containing information about the EP has been identified, an optimum digital filter can be defined, which will selectively remove the shaded part of the frequency range from subsequent samples of the EP.

The optimum digital filter is different for each kind of EP, varies from patient to patient depending upon the individual pathology, response to medication and anesthetics and can change within an individual patient as a result of decompression or other intraoperative factors. Thus, the optimum digital filter cannot be specified beforehand and therefore preselected values cannot be used. In order to use this technique properly, a system is needed which permits rapid identification of the optimum digital filter in any situation.

INTRAOPERATIVE MONITORING
UTILIZATION
OF THE BRAIN STATE ANALYZER
AT UNIVERSITY HOSPITAL,
N.Y.U.M.C.
(6/82 - 6/85)

Brainstem / Posterior Fossa	42
Pituitary Tumors	17
Acoustic Neuromas	81
Spinal Cord Tumors	95
Aneurysms / AVMs	34
Neuroembolizations	25
Coma Evaluations	23
Cardiopulmonary Bypass Surgery	61
Post-Operative Evaluations	30

Fig. 3

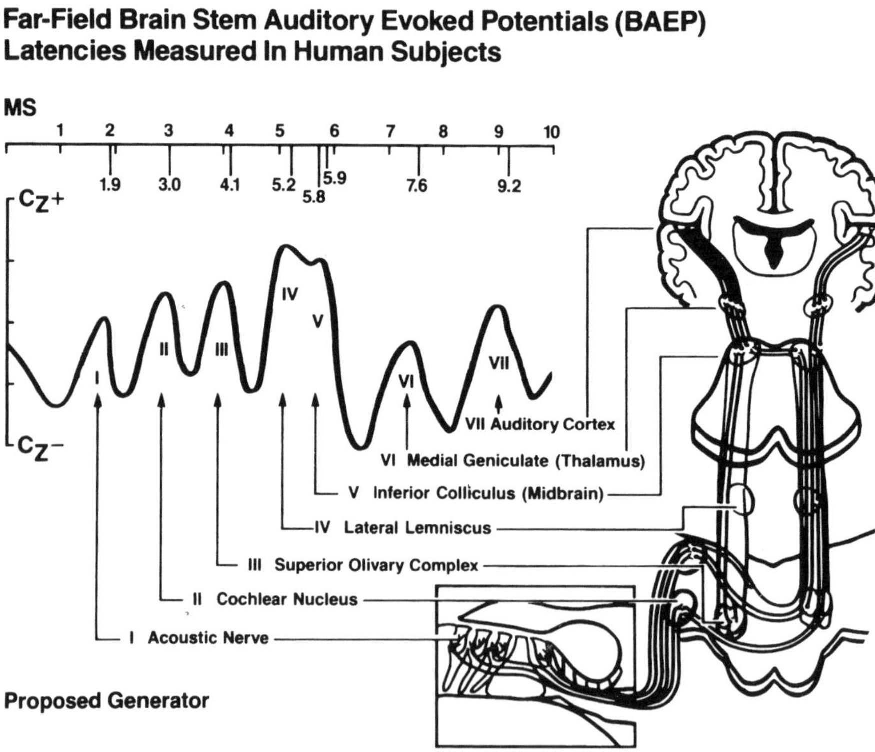

Fig. 4

Such a system has been programmed into an instrument which we call the Brain State Analyzer, or BSA. In the last 3 years, we have evaluated the BSA in large groups of patients undergoing many different surgical procedures, as shown in Fig. 3. The removal of noise from the EP by optimum digital filtering is so efficient that in many of these operations, reliable successive measurements, which we call "updates" could usually be obtained every 5–10 s.

Fig. 4 shows the neuroanatomical structures currently believed to be the generators of the successive peaks of the brain stem auditory evoked response, or BAER. In normal subjects, the BAER displays 7 peaks at latencies which are very reproducible from person to person. The first 5 peaks are believed to reflect the successive activation of the acoustic nerve, cochlear nucleus, superior olivary complex, lateral lemniscus, and inferior colliculus, by the afferent volley caused by a click stimulus.

Fig. 5 shows the screen of the BSA displaying BAERs while monitoring resection of an acoustic neuroma. In this and all BSA screens which will be shown, the following convention is used:

The top waveshape is a "reference baseline" obtained by averaging 2048 EPs after optimum digital filtering, usually collected early in the procedure. The second waveshape is also based upon 2048 EPs and represents the most recent or "current" baseline. The bottom 3 waveshapes represent successive updates based upon much smaller samples which reflect the present functional condition of the monitored system.

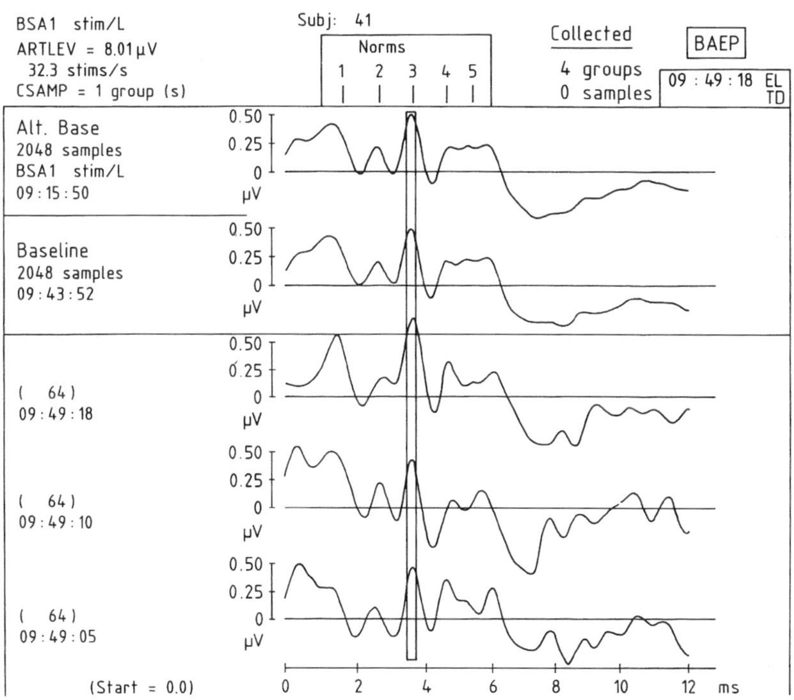

Fig. 5

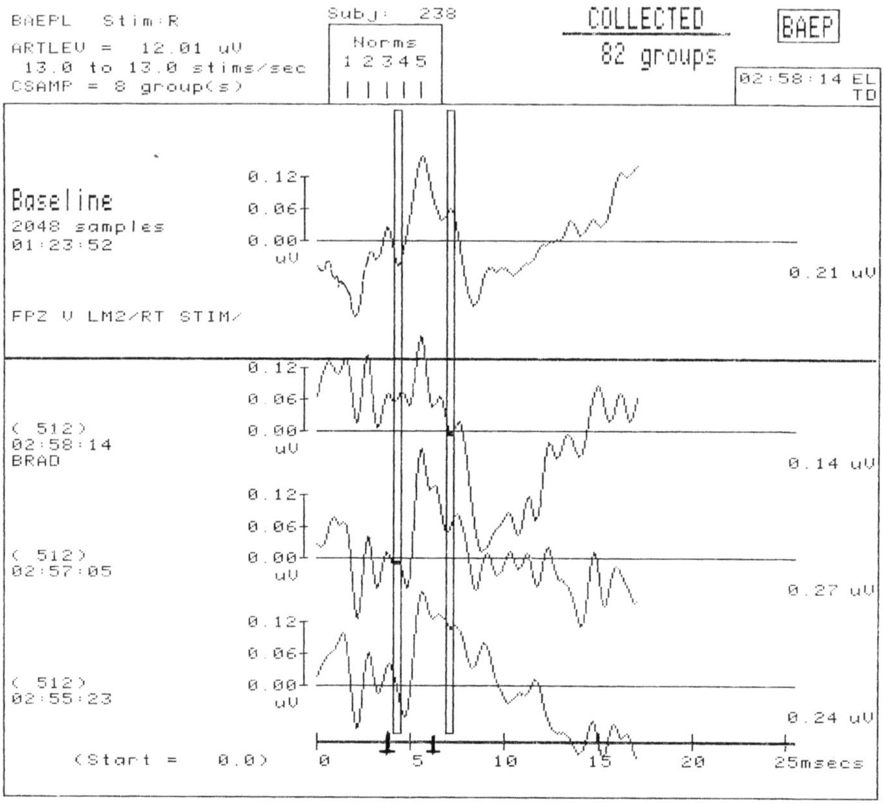

Fig. 6

The 3rd waveshape is the most recent update, the 4th preceded the 3rd and the 5th preceded the 4th. These permit the recent history of functional changes to be seen at a glance. Each update in this example was based upon 64 samples, and new updates were obtained every 6 seconds. The vertical lines show the 95% confidence interval for the latency of the third positive peak. Note the extreme stability of the updates at this time.

Fig. 6 illustrates the sensitivity of the BAER to the placement of a retractor which caused severe EP depression followed by bradycardia. The update interval was slower because of detection of unacceptable electrical noise in the OR by an automatic artifact rejection program in the BSA.

Because of the efficient noise removal, this method permits a great increase in sensitivity so that smaller EP components can reliably be monitored. Fig. 7 shows a BSA screen recorded during acoustic neuroma surgery, using the BSA as a conventional average response computer. Even though the updates are based upon 1024 samples, environmental noise in the OR completely obscures the morphology of the EP waveshape.

In Fig. 8 are the same data displayed with optimum digital filtering. The morphology of the EP waveshape is now very clear. Notice the negative complex between 12

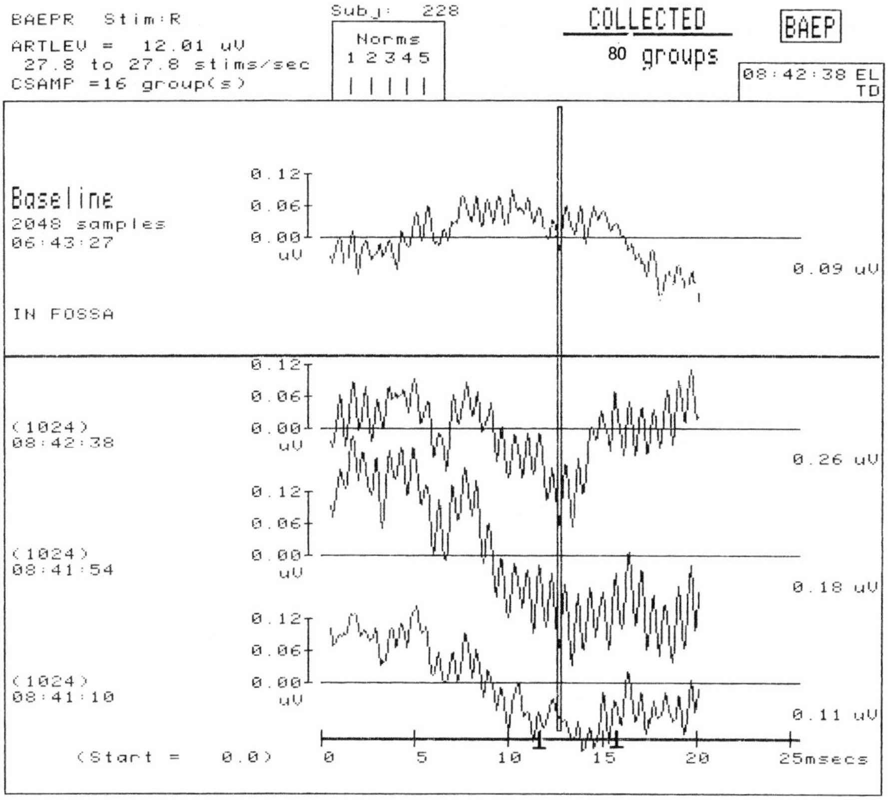
BAEPR Stim:R
ARTLEV = 12.01 uV
27.8 to 27.8 stims/sec
CSAMP =16 group(s)
Subj: 228
Norms
1 2 3 4 5
COLLECTED
80 groups
BAEP
08:42:38 EL
TD
Baseline
2048 samples
06:43:27
IN FOSSA
0.09 uV
(1024)
08:42:38
0.26 uV
(1024)
08:41:54
0.18 uV
(1024)
08:41:10
0.11 uV
(Start = 0.0)
25msecs

Fig. 7

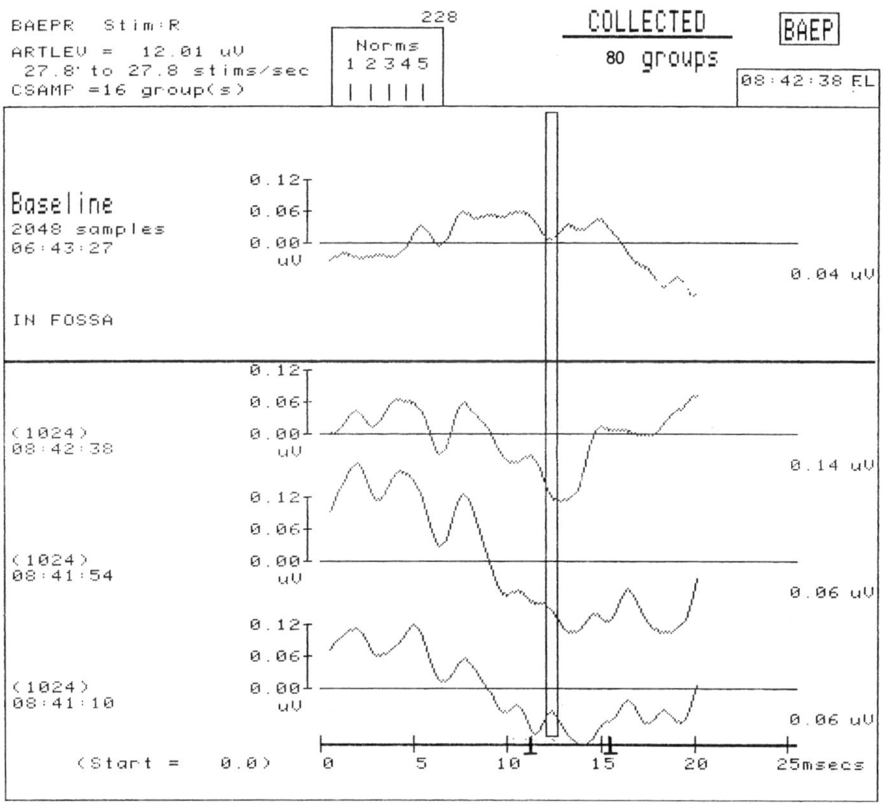
BAEPR Stim:R
ARTLEV = 12.01 uV
27.8 to 27.8 stims/sec
CSAMP =16 group(s)
228
Norms
1 2 3 4 5
COLLECTED
80 groups
BAEP
08:42:38 EL
Baseline
2048 samples
06:43:27
IN FOSSA
0.04 uV
(1024)
08:42:38
0.14 uV
(1024)
08:41:54
0.06 uV
(1024)
08:41:10
0.06 uV
(Start = 0.0)
25msecs

Fig. 8

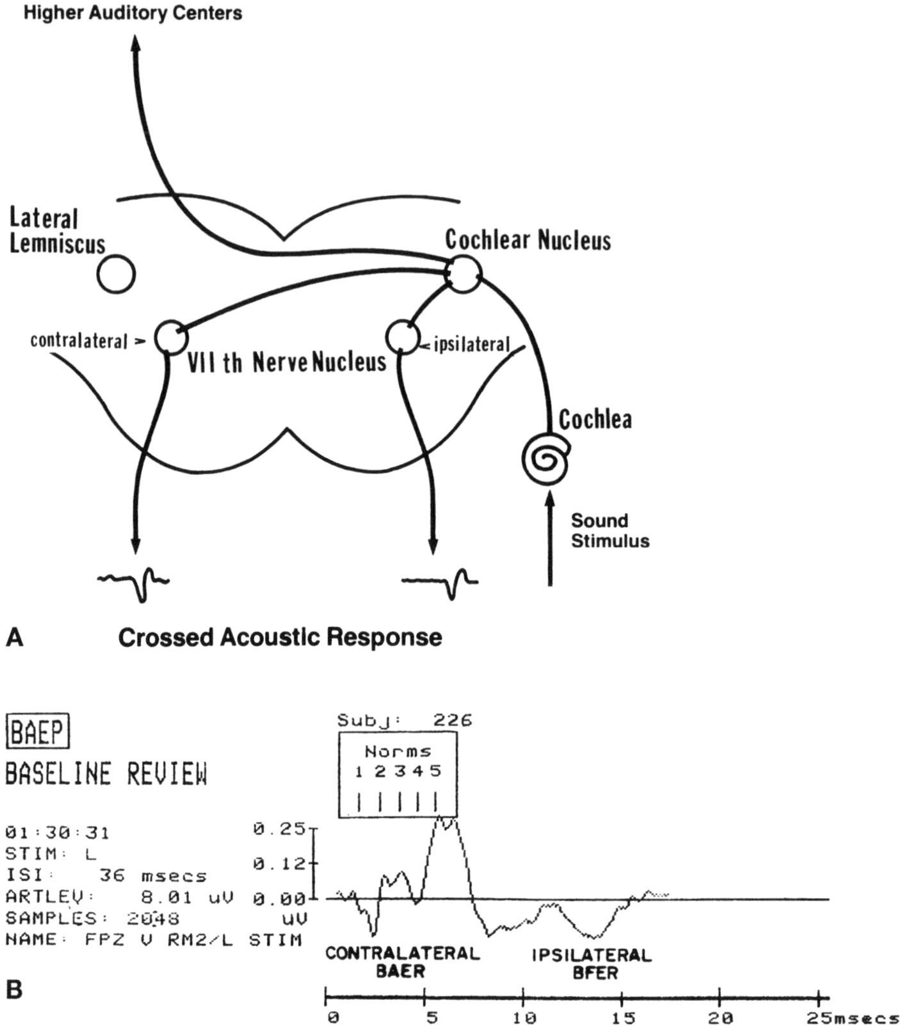

Fig. 9A, B

and 15 milliseconds, highlighted by the vertical confidence interval. We routinely use this complex, often as small as 0.03 microvolts in amplitude, to monitor the integrity of the VIIth cranial nerve, especially during acoustic neuroma surgery. We call it the brain stem facial evoked response, or BFER.

The BFER is based upon the crossed acoustic reflex, or post-auricular reflex. This reflex activates the motor outflow of the facial nerve in response to auditory stimulation, and is the vestigial remnant in man of the pinna reflex (see Fig. 9A). The reflex can be elicited ipsi- or contralateral to the stimulated ear. Contralateral activation is particularly advantageous when hearing is compromised on the involved side, since it permits the facial nerve on the involved side to be monitored using input from the intact side, which is always available.

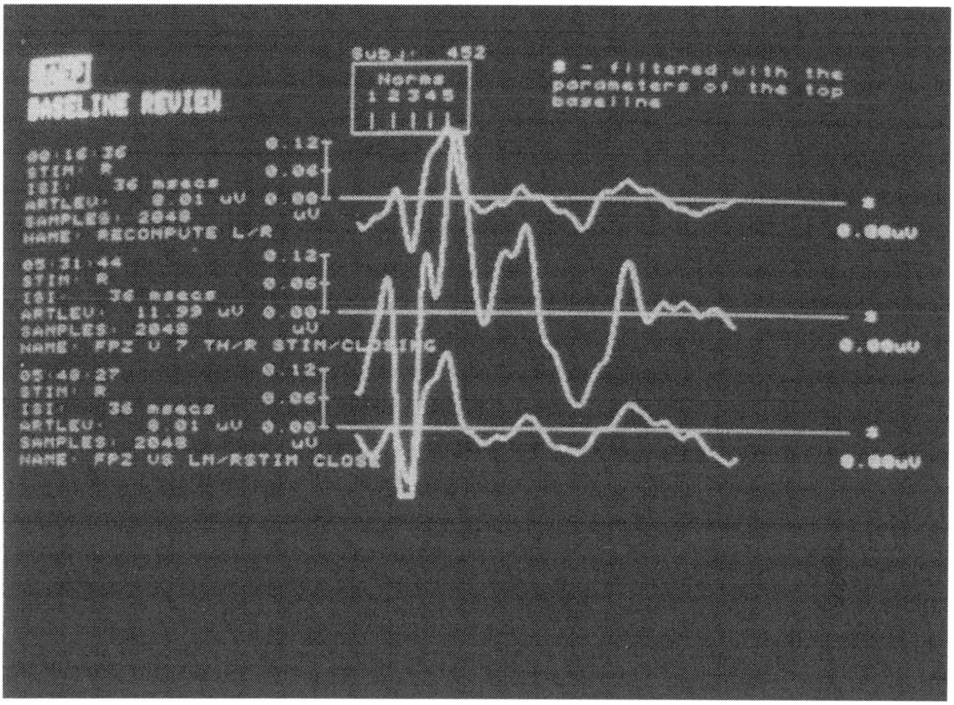

Fig. 10

The resulting EP waveshape seen on Fig. 9B reflects the contralateral BAER from 0 to 6 or 8 milliseconds and the ipsilateral BFER in the interval between 10 and 15 milliseconds. A further advantage of this technique is that the contralateral BAER serves as a constant indicator of general brain stem condition, reflecting systemic or nonspecific influences not restricted to the actual surgical site.

We have used this method to monitor over 50 acoustic neuroma and other operations in which the facial nerve was at risk. The surgeons who performed most of these procedures, Drs. Joseph Ransohoff and Noel Cohen, have repeatedly confirmed the correlation between changes in the BFER complex between 12 and 14 milliseconds and intraoperative manipulations of the VIIth cranial nerve. Because of the practical utility of a continuous electrophysiological monitor of facial nerve function, we considered it important to obtain a direct demonstration that the BFER complex arose from the facial nerve. On 2 occasions, we have had the opportunity to record directly from the facial nerve during surgery, with essentially identical observations in both cases.

One of these cases is illustrated in Fig. 10. The top and bottom waveshapes represent baseline samples of the BFER at the beginning of an acoustic neuroma operation and at the end of the resection, 5½ hours later. These waveshapes represent the optimally filtered average of 2048 responses, recorded between a frontal electrode at FPZ and an electrode on the ipsilateral mastoid. The middle waveshape shows the waveshape obtained when the electrode on the mastoid is replaced by one placed

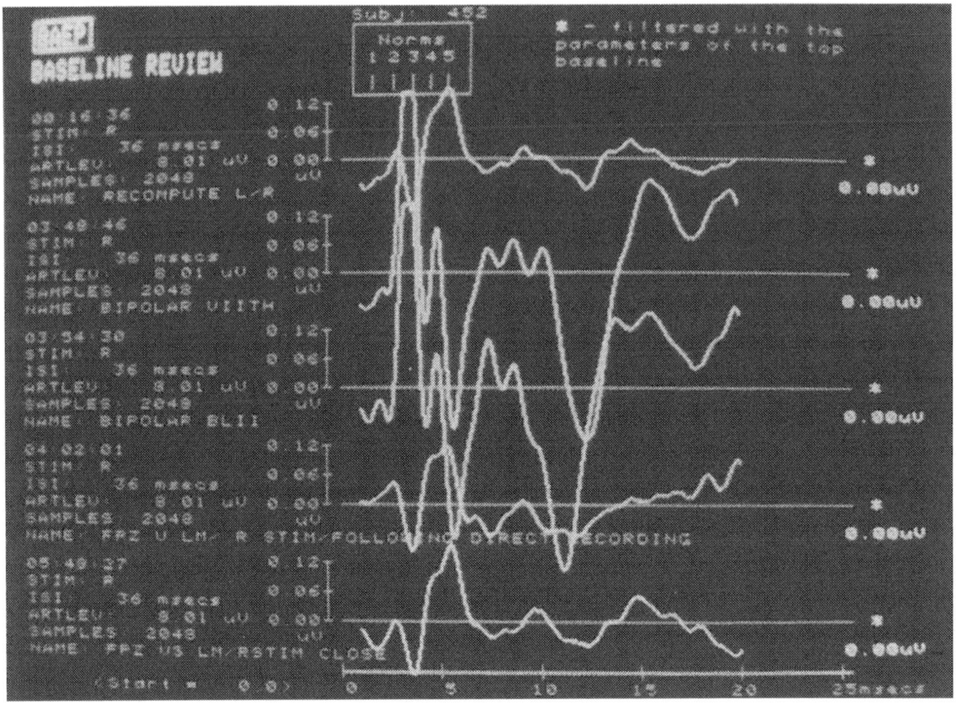
BSEP
BASELINE REVIEW
Subj: 452
Norms
1 2 3 4 5
* - filtered with the parameters of the top baseline
STIM: R
ISI: 36 msecs
ARTLEV: 8.01 uV
SAMPLES: 2048
NAME: RECOMPUTE L/R
NAME: BIPOLAR VIITH
NAME: BIPOLAR BLII
NAME: FPZ V LM/ R STIM/FOLLOWING DIRECT RECORDING
NAME: FPZ VS LM/RSTIM CLOSE
0.12
0.06
0.00 uV
0.00uV
<Start = 0.0>
0 5 10 15 20 25msecs

Fig. 11

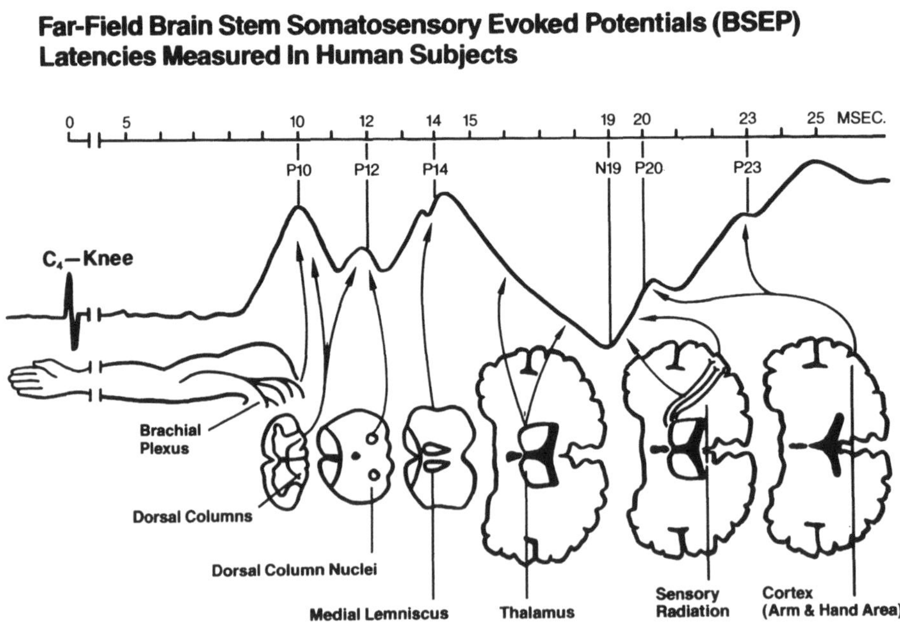
Far-Field Brain Stem Somatosensory Evoked Potentials (BSEP)
Latencies Measured In Human Subjects
0 5 10 12 14 15 19 20 23 25 MSEC.
P10
P12
P14
N19
P20
P23
C4–Knee
Brachial Plexus
Dorsal Columns
Dorsal Column Nuclei
Medial Lemniscus
Thalamus
Sensory Radiation
Cortex (Arm & Hand Area)

Fig. 12

directly on the VIIth nerve. The 3 waveshapes are very similar. Note the enormous increase in the negative going potential which coincides with the BFER peak complex between 12 and 14 milliseconds.

Bipolar recordings, with both electrodes directly on the VIIth nerve, are illustrated in Fig. 11 (waves 2 and 3). Waves 1, 4 and 5 show the far-field BFER before and after direct bipolar recording from the VIIth nerve. The electrodes were placed on the facial nerve in the internal auditory canal and on its horizontal segment. Both sites were well away from the trigeminal cranial nerve. Preoperative evoked EMG could not detect any electrical activity on the side of the tumor in this clinically intact patient, in contrast to the clear BFER. We consider these data to provide unequivocal evidence that the generator of the negative part of the BFER complex is the facial nerve. The origin of the earlier positive components is still unclear, but may be the nucleus of the VIIth nerve.

The brain stem somatosensory evoked response, or BSER, is the EP measure which we have found useful in the widest range of procedures: neurosurgery in the brain stem, posterior fossa, and spinal cord; clipping of aneurysms and neuroembolization of arteriovenous malformations. Fig. 12 shows the presumed anatomical origins of the BSER, with successive components believed to reflect the activation of the dorsal column nuclei, medial lemniscus, thalamus, sensory radiation, and the first

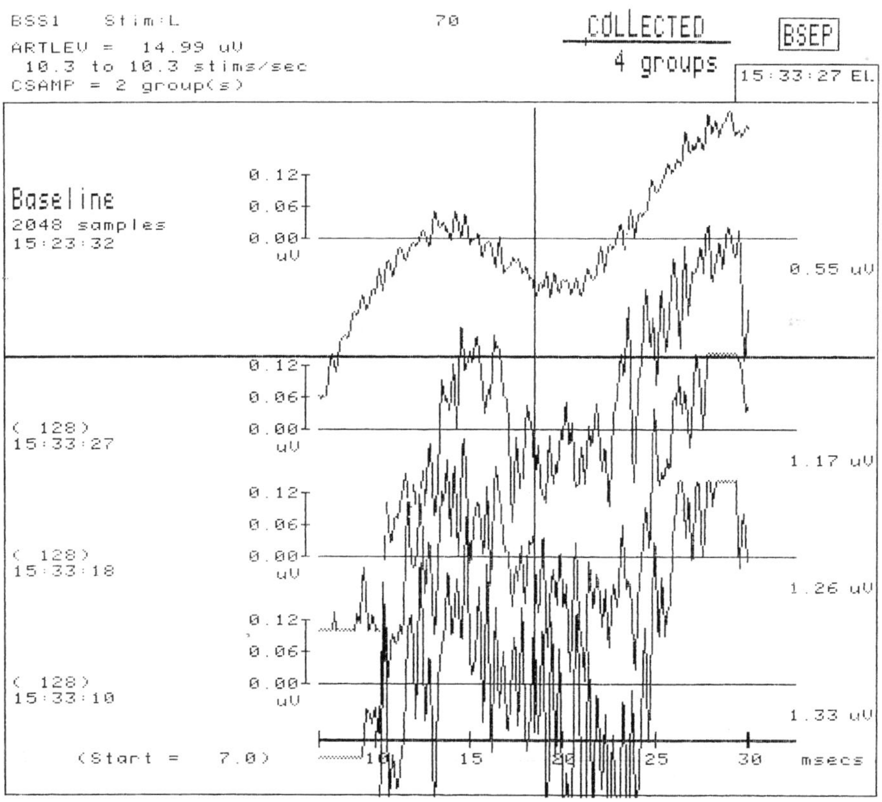

Fig. 13

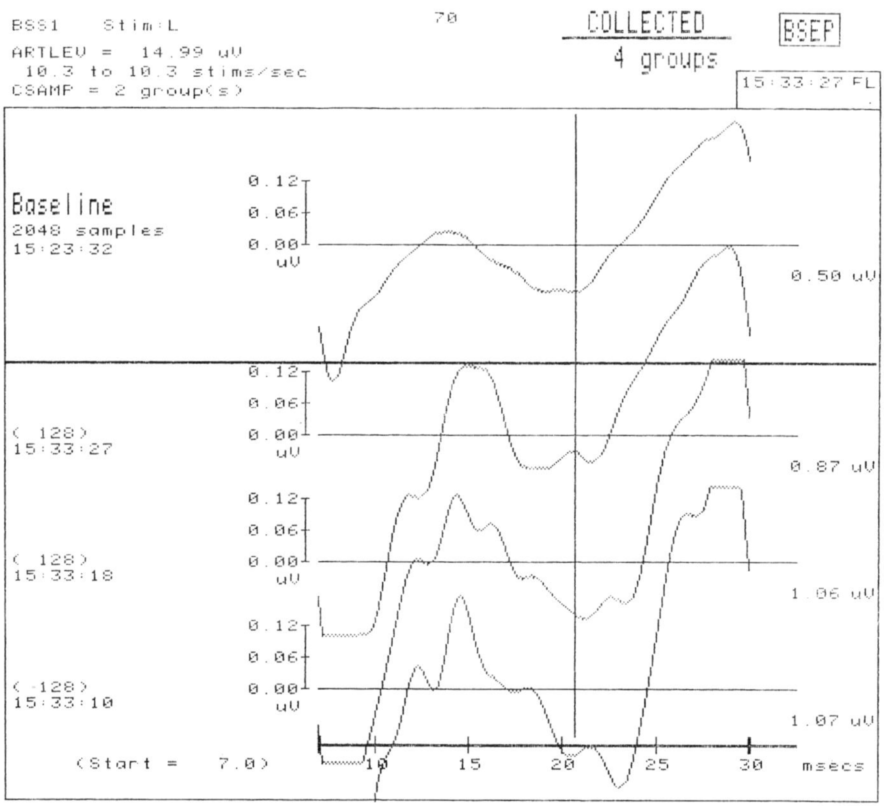

Fig. 14

cortical synapses. The latency between the positive medial lemniscus peak, normally at 14 milliseconds, and the first cortical synapse, normally at 23 milliseconds, is sometimes referred to as the "central conduction time". In our experience, amplitude decrements in P_{14} and N_{19} are often dissociated from each other and we believe that these components may reflect relatively pauci- and multi-synaptic pathways which conduct at different velocities, rather than the different anatomical structures proposed in this diagram. Amplitude decrements usually precede latency increases and, in our opinion, are more sensitive indicators of tissue distress.

Fig. 13 shows the BSER during resection of an intra-medullary spinal tumor, using the BSA as a conventional averager. The upper 2 waveshapes represent reference baselines computed from 2048 samples, while the lower waveshapes show the last 3 updates based on 128 samples each, with an update time of 8–10 s. Noise almost completely obscures the EP waveshape.

Fig. 14 shows the same BSER data, using the BSA with optimum digital filtering. Notice the latency decrease of the negative going component inside the vertical confidence interval of the second baseline compared with the first baseline. This reflects the rapid increase in transmission speed with decompression due to debulking of the tumor, and is often prognostic of post-operative clinical improvement. Also notice

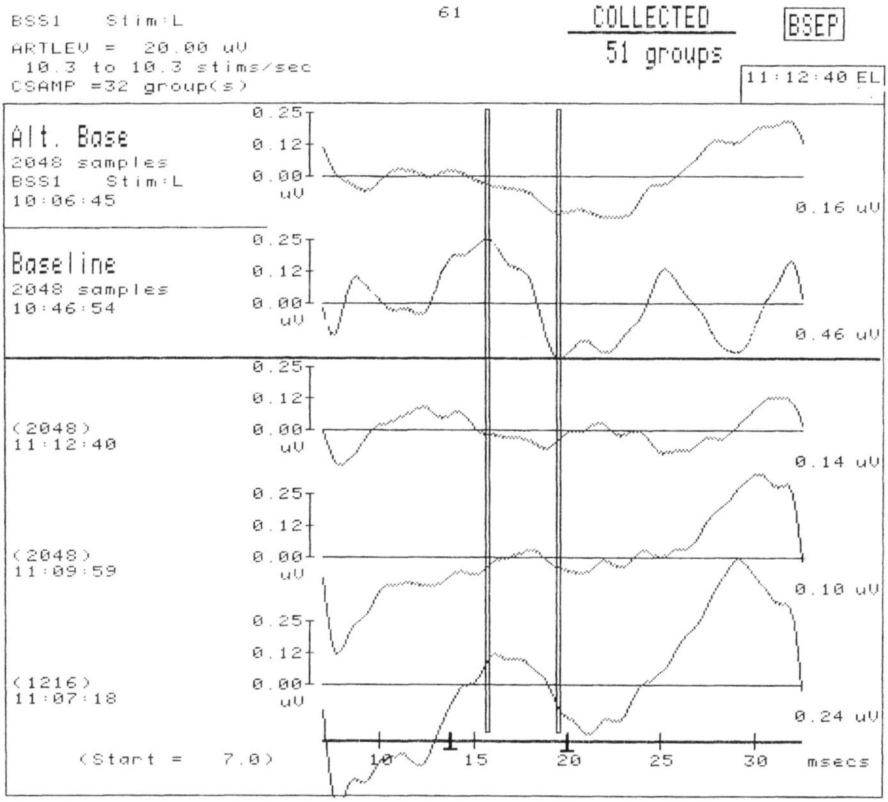

Fig. 15

the shorter latency of the most recent update in the 3rd waveshape compared with the 2 preceding updates below. This reflects rapid recovery from a transient slowing caused by surgical manipulations. Information about such transient changes constitutes feedback which often helps the surgeon to evaluate the current situation.

Fig. 15 also illustrates BSER changes during resection of an intramedullary tumor in a child. Notice the marked increase in amplitude and decrease in latency of the BSER in the second baseline compared with the first, taken an hour earlier. This BSER improvement correlated with striking post-operative clinical improvement in this child, who was severely impaired prior to surgery. The updates below show the drastic decrease in amplitude and increase in latency with which the BSER reflects the functional status of the tissue. This is particularly useful when using instruments such as the laser, which deprive the surgeon of familiar tactile feedback. Heat generated by the laser seems to accumulate as some constituents are volatilized, as if the specific heat of tissue residues was increasing. Evaluating the need for pauses and irrigation is aided by on-line functional monitoring as illustrated here. Cavitron effects can similarly be better evaluated when monitoring is available.

Fig. 16 illustrates stability of the BSER as the surgeon exposes an aneurysm of the middle cerebral artery.

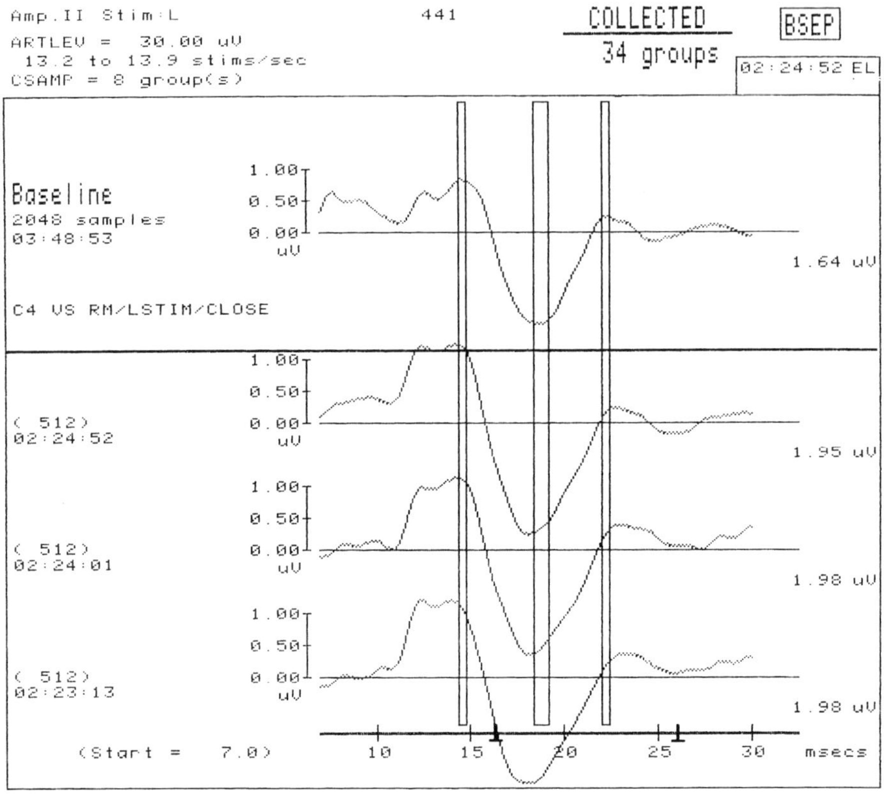

Amp.II Stim:L
ARTLEV = 30.00 uV
13.2 to 13.9 stims/sec
CSAMP = 8 group(s)
441
COLLECTED
34 groups
BSEP
02:24:52 EL
Baseline
2048 samples
03:48:53
C4 VS RM/LSTIM/CLOSE
1.00
0.50
0.00
uV
1.64 uV
(512)
02:24:52
1.95 uV
(512)
02:24:01
1.98 uV
(512)
02:23:13
1.98 uV
(Start = 7.0)
10
15
20
25
30
msecs

Fig.16

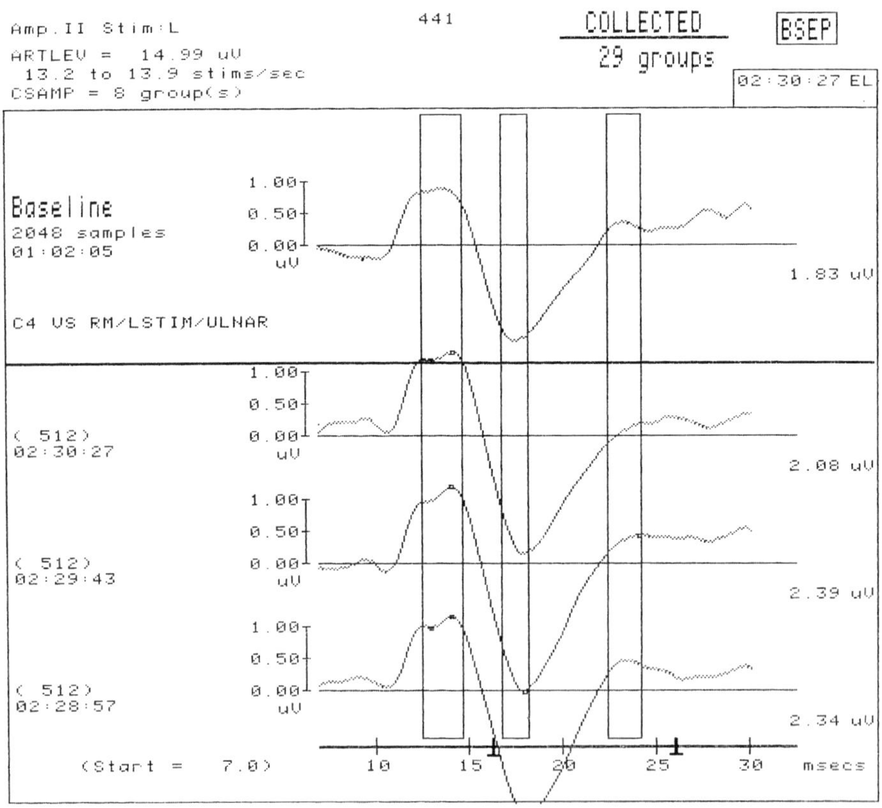

Amp.II Stim:L
ARTLEV = 14.99 uV
13.2 to 13.9 stims/sec
CSAMP = 8 group(s)
441
COLLECTED
29 groups
BSEP
02:30:27 EL
Baseline
2048 samples
01:02:05
C4 VS RM/LSTIM/ULNAR
1.00
0.50
0.00
uV
1.83 uV
(512)
02:30:27
2.08 uV
(512)
02:29:43
2.39 uV
(512)
02:28:57
2.34 uV
(Start = 7.0)
10
15
20
25
30
msecs

Fig.17

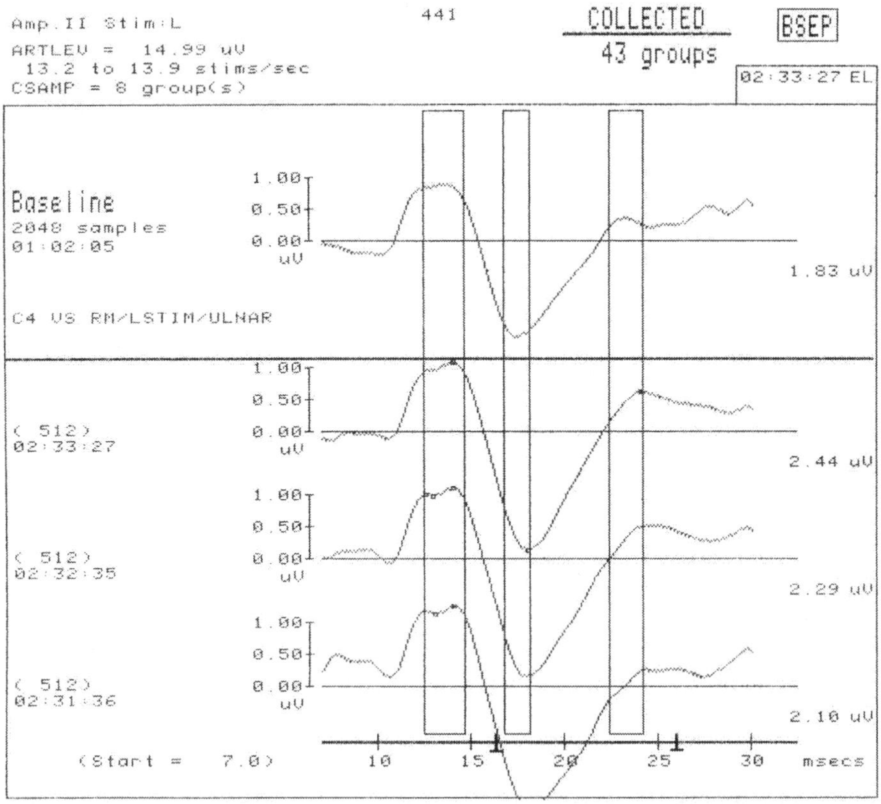

Amp.II Stim:L
441
COLLECTED
43 groups
BSEP
ARTLEV = 14.99 uV
13.2 to 13.9 stims/sec
CSAMP = 8 group(s)
02:33:27 EL
Baseline
2048 samples
01:02:05
1.00
0.50
0.00
uV
1.83 uV
C4 VS RM/LSTIM/ULNAR
(512)
02:33:27
2.44 uV
(512)
02:32:35
2.29 uV
(512)
02:31:36
2.10 uV
(Start = 7.0)
10
15
20
25
30
msecs

Fig. 18

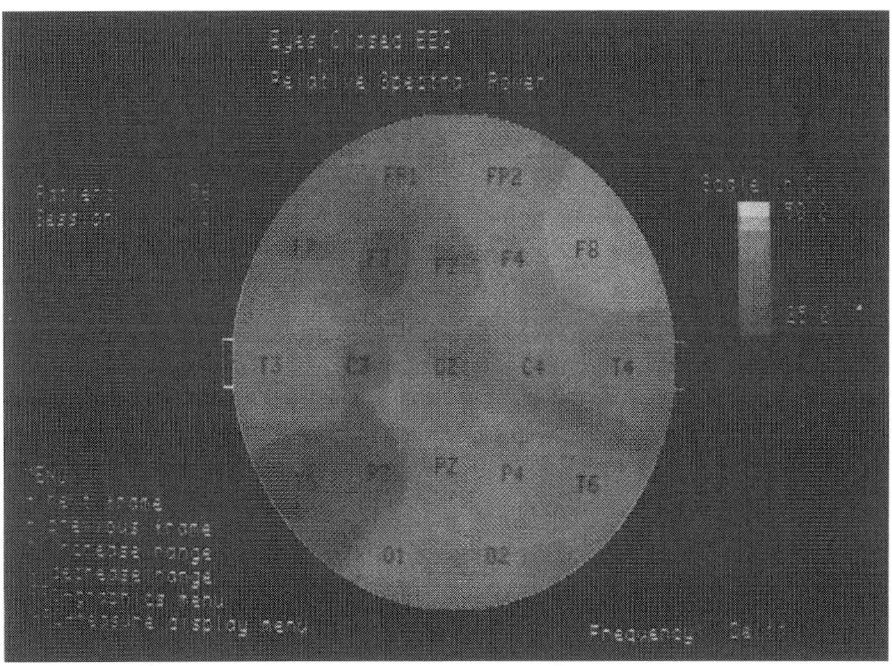

Eyes Closed EEG
Relative Spectral Power
FP1
FP2
F3
F4
F8
T3
C4
T4
PZ
P4
T6
O1
O2
Frequency:

Fig. 19

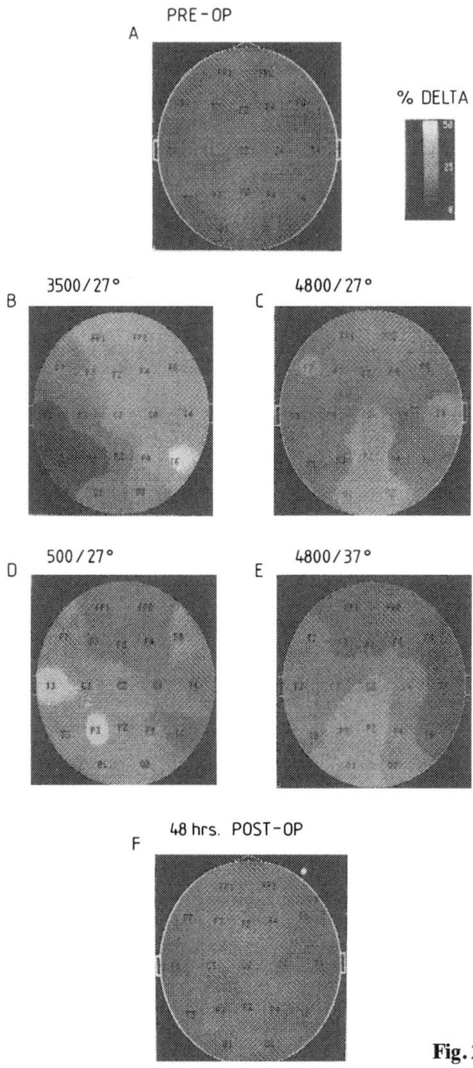

Fig. 20

Fig. 17 shows an increase in central conduction time when a temporary clip was placed on the aneurysm at 2:28:42. Notice that the positive peak in the most recent update at 2:30:27, the third waveshape, has moved significantly beyond the confidence interval indicated by the vertical lines around 23 milliseconds.

Fig. 18 illustrates the steady return to the previous latency beginning when the temporary clip was removed at 2:31:36 (wave 5). By 2:33:27, the central conduction time has returned to its previous speed, as seen in the 3rd waveshape. In this way, the surgeon can be helped to evaluate the cerebral ischemia caused by alternative clip placements, and can be guided to find a placement with minimal functional consequences. We have found the BSEP of similar utility in evaluation of spinal cord ischemia during resection of an aneurysm of the abdominal aorta.

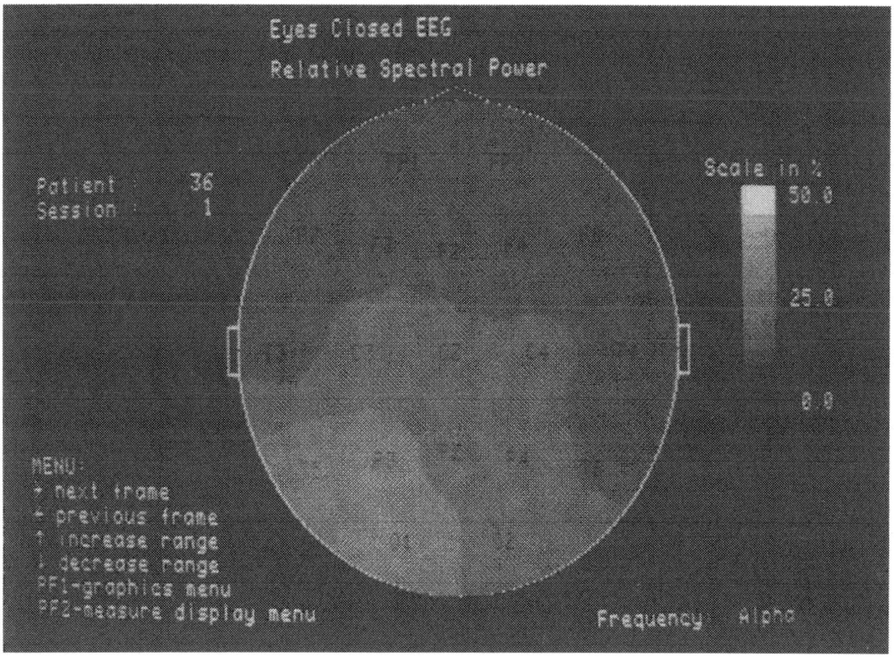

Fig. 21

We have successfully used optimum digital filtering with visual cortical evoked potentials to monitor resection of pituitary and sub-orbital tumors and with neuro-embolizations of the posterior cerebral artery, which cannot be illustrated in the time available here. I prefer instead to turn to a different monitoring technique also available in the BSA, which is the on-line construction of topographic field maps of the frequency composition of the EEG or the distribution of EP voltage as a function of latency. Such maps can be used for many purposes.

In collaborative studies of patients with cerebrovascular disease we have shown that quantitative measures of the EEG frequency spectrum are very much more sensitive to regional cerebral blood flow than Xe^{133} measurements. Our data suggest that it may be possible to identify the stroke-prone patient. These findings have many applications of clinical utility.

Fig. 19 shows the topographic map of the distribution of slow EEG activity in the delta frequency band, in a patient who suffered a right middle cerebral artery infarct during surgery involving cardiopulmonary bypass, or CPB. One is looking down at the head face upward. The brighter the map, the poorer the regional cerebral blood flow. The patient was left hemiplegic and in coma. The infarct was later confirmed by CT scan.

Fig. 20 is a composite of such maps obtained while monitoring a coronary artery bypass procedure. The top head shows the preoperative anesthetized map of cerebral blood flow. Heads B and D illustrate the onset of cerebral ischemia with low pump volume, while C and E show the reversal of signs of cerebral distress when flow was increased. The bottom map was obtained postop.

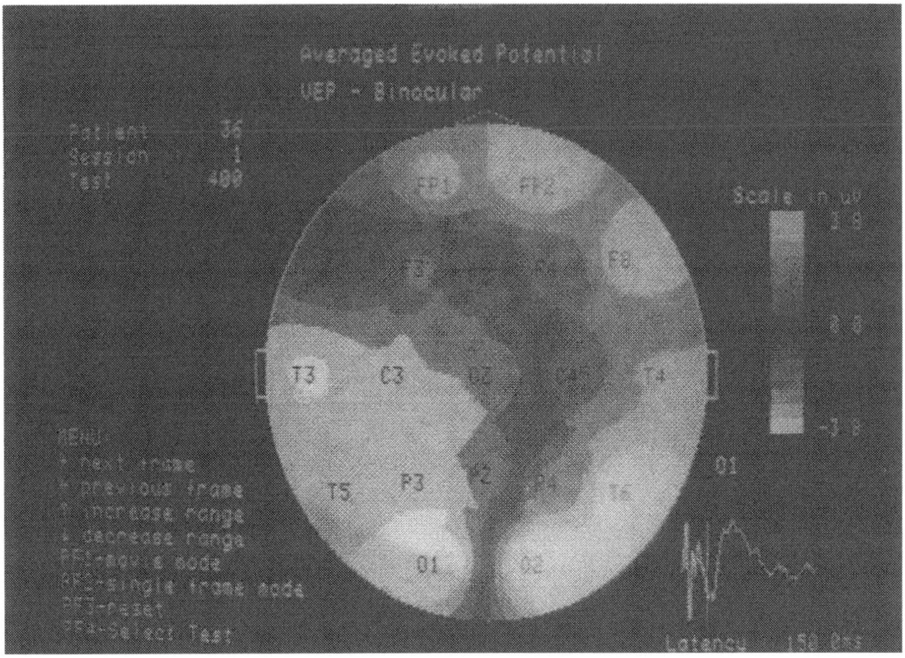
Averaged Evoked Potential
VEP - Binocular
Patient 36
Session 1
Test 400
FP1
FP2
F3
F4
F8
T3
C3
CZ
C4
T4
T5
P3
PZ
P4
T6
O1
O2
Scale in uV
3.8
0.0
-3.8
O1
MENU
next frame
previous frame
increase range
decrease range
PF1-movie mode
PF2-single frame mode
PF3-reset
PF4-Select Test
Latency 150.0ms

Fig. 22

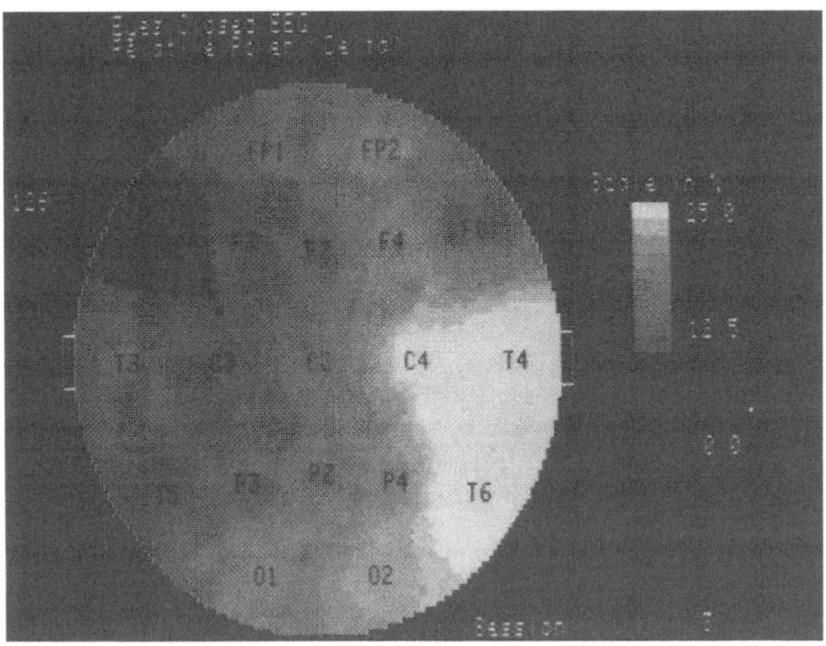
FP1
FP2
F4
T3
C4
T4
P4
T6
O1
O2
25.0
12.5
0.0
Session

Fig. 23

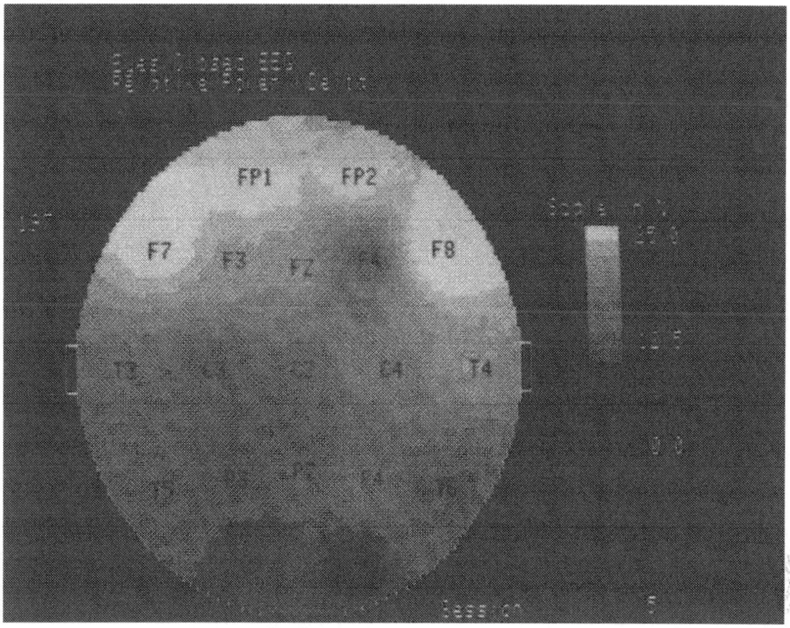
FP1
FP2
F7
F3
FZ
F4
F8
T3
C4
T4
T5
Session
5

Fig. 24

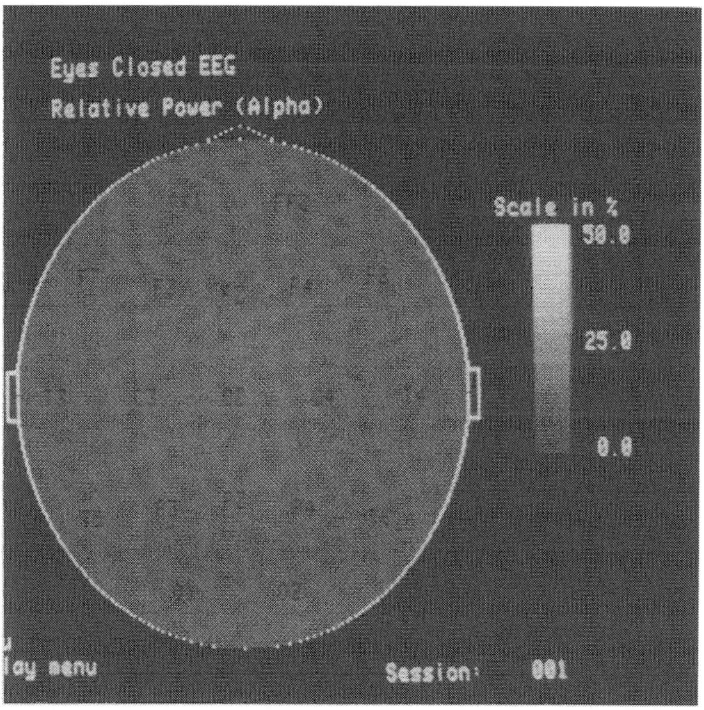
Eyes Closed EEG
Relative Power (Alpha)
Scale in %
50.0
25.0
0.0
Session: 001

Fig. 25

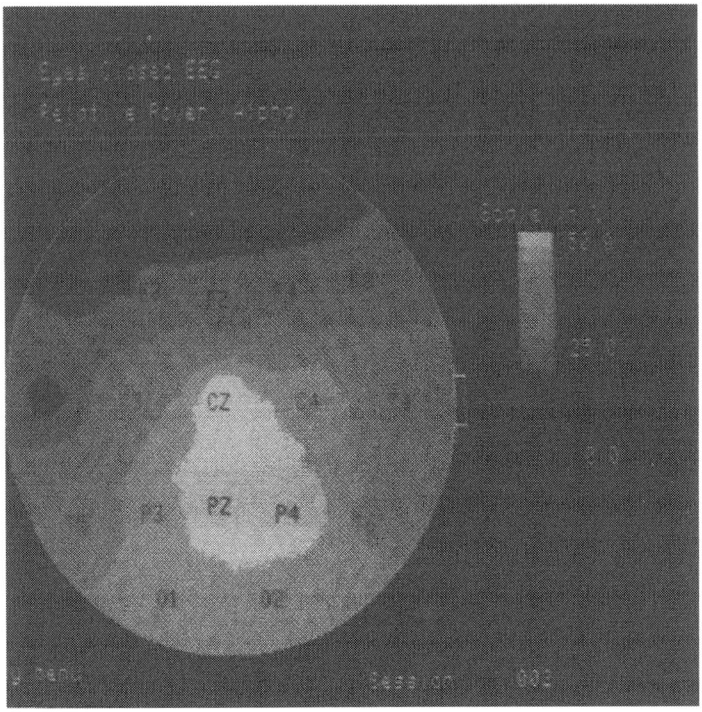

Fig. 26

Fig. 21 is a map of asymmetrical alpha distribution in a patient after CPB. The alpha map is normally symmetrical, with a posterior maximum. This patient was subsequently discovered to be left hemianopic, although the CT scan was negative.

Fig. 22 is a map of the electrical field of the visual evoked potential in the same patient 150 milliseconds after a binocular full field flash was presented. Positive voltage is indicated by shades of red, while negative is indicated by blue. The extreme posterior asymmetry confirms that the thalamo-cortical projection to the right occipital cortex is dysfunctional.

Fig. 23 is a delta field map of a patient undergoing carotid endarterectomy. The right carotid in this patient was 100% stenotic, while the left carotid was 75% occluded. This map was obtained after test clamping of the left carotid. Notice the signs of severe cerebral steal on the right side.

Similar contralateral cerebral ischemia can be seen in the map shown in Fig. 24, immediately after balloon inflation prior to embolization of a left carotid AVM.

We have found quantitative EEG and EP analyses very useful in the evaluation of coma. Fig. 25 is the alpha frequency map of a patient with "locked-in" syndrome, after a vertebral subarachnoid hemorrhage and vasospasm. Notice the uniform absence of alpha activity.

A cognitive task requiring mental imagery was then presented to this apparently comatose patient. Fig. 26 notice the marked alpha activation in posterior regions. We believe that such tests may reveal the potential capabilities of non-reactive patients.

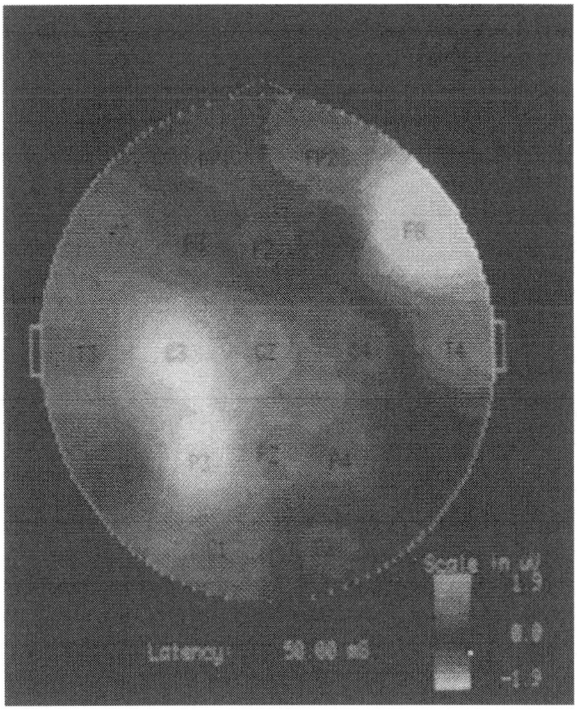
FP2
F8
F3
FZ
T3
C3
CZ
C4
T4
P3
PZ
P4
Scale in uV
1.9
0.0
-1.9
Latency: 50.00 mS

Fig. 27

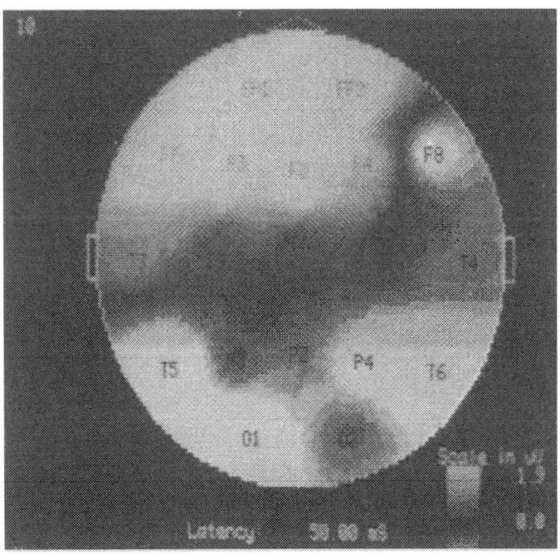
10
FP2
F8
F3
FZ
F4
T4
T5
PZ
P4
T6
O1
Scale in uV
1.9
0.0
Latency: 50.00 mS

Fig. 28

Fig. 27 shows the field map evoked 50 milliseconds after a meaningless click, in a patient with expressive aphasia after subarachnoid hemorrhage of the left middle cerebral artery.

Fig. 28 is the field map of the same patient, 50 milliseconds after presentation of the click, while engaged in a task requiring pointing to the correct name of a visual shape. This technique is referred to as the "non-contingent probe" method. Notice the marked change in the left fronto-temporal regions. This patient is now receiving speech therapy.

In this paper, I have presented an overview of our experience with quantitative EP and EEG analysis, as implemented in the Brain State Analyzer. We have found that intra-operative monitoring of cortical or brain stem EPs, using optimum digital filtering, and of quantitative EEG, using topographic mapping, is extremely helpful to the surgeon. On one hand, it reassures the surgeon, telling him that it is safe to proceed further. On the other hand, it provides clear early warning signals of imminent danger. In our opinion, such monitoring significantly reduces unexpected post-operative neurological deficits and facilitates rapid recovery.

Further, we have found EP and EEG analysis, particularly in combination with topographic mapping, to be extremely useful in assessment of the functional condition of impaired patients. The insights provided by such assessment are expected to be particularly useful in pharmacological interventions and in the general management of such patients.

The Monitoring of the Brain Stem and the Auditory Function in Neurosurgery

M. J. Soulier, Y. Lazorthes, B. Fraysse, and M. Vincent

Brain stem auditory evoked potentials (BAEPs) permit the recording of the auditory systems response to physiological stimulation.

BAEPs analyze this auditory response during the first milliseconds after stimulation and they currently constitute an essential examination in audiological and neurological diagnosis.

Because these early BAEPs provide information about the functional integrity of the auditory nervous structures, and about the anatomical regions these structures run through, (Hashimoto et al. 1981; Moller and Jannetta 1983), recording BAEPs appears advantageous during neurological surgery involving risk to these nervous structures.

Recording BAEPs has become a classical kind of scanning carried out in all neurophysiology units. However, these intra-operative recordings are made more difficult by the electrical environment which constitutes the operating room. (Grundy 1982; Levine et al. 1984; Raudzens and Shetter 1982; Zappulla et al. 1984).

The neurophysiologist is thus confronted with news parameters and different information which must be analyzed and quantified so as to give useful information as quickly as possible to the surgeon.

Material and Methods

Material

We find it necessary at present to distinguish between two kinds of data acquisition methods of averaging evoked potentials, for functional sensori-motor examinations, which differ in their manner of recording the signal and their speed.

The first technique, by means of *digital filters,* allows more immediate visualization of biological signals every 5–10 s. The almost instantaneous updated delivery of signals and the confidence intervals of the main parameters make it possible to give the neurosurgeon data on the functioning of the brain stem and the tested sensory function synchronous with surgery.

The second possibility of monitoring the intra-operational nervous sensory functions is use of *analog filters.*

Indeed, an action potential, whether it be auditory, somesthesic or visual, has a very weak amplitude of a few hundred nanovolts or less, and is submerged in a much more intense background noise, consisting of cortical activity and neighboring regions muscle potential.

Averaging consists of repeating a physiological stimulation a certain number of times (1000 or 2000) at high speed, 10 to 20 times per second, for instance, for sound stimulation, which results in the improvement of the signal to noise ratio and the extraction of the action potential signal.

Initially we used a classic averaging device with analog filters (Medelec MS6) with 2 traces, and plotting on a table XY.

We recently acquired a larger more versatile computer system, (Nicolet Pathfinder), with 4 channels and 16 traces, with independent memories, which can be simultaneously run on a viewing screen. This allows for rapid comparison of different operating times.

The computer processing unit permits the automatic acquisition and storage averages, with the possibility of post-operative statistical calculations.

This device comes specially adapted for monitoring with a program of digital filters, which allows the recognition of a response, in 15 to 20 second, after 300 to 400 stimulations, averaging which is not the real time, but which provides information sufficiently immediate for the neurosurgeon.

Methods

For the 26 examinations performed for surgery on the posterior fossa, our recording parameters were identical.

The analysis time is 10–20 ms; the passing band is 160 Hz to 3,2 kHz and the amplifier gain 10×5. The sound stimulation is an unfiltred click of 120 microseconds duration with a frequency of 20 times per second.

The transducer is a TDH 39, and the stimulus intensity is 90 or 100 dB, peak equivalent SPL depending on the patient's auditory threshold.

A systematic controlateral masking is used.

To get a better baseline in such a hostile place as an operating room, with many electronic devices present, we plug our reception channel into a transformer and isolated amplifier, and we also insulate the patients head, since most or our patients are operated in a sitting position.

Our protocol consists of taking recordings as continuous as possible, each averaging lasting 1 to 2 minutes with the first apparatus and less than one minute with the second.

Results

Our results concern 26 recordings carried out during posterior fossa operations. These included:

- 13 neurinomas,
- 2 meningiomas,
- 3 microvascular decompressions,
- 3 aneurims,

- 1 cerebellar medullobastoma,
- 1 brain stem tumor,
- 1 occipital aperture tumor,
- 2 vestibular nerve sections.

We have chosen to explain in detail 3 cases which seem to us to illustrate these different recordings well.

Case No. 1

A 23-year-old woman presented a progressive deafness associated with vertigo, known for 10 years, recently aggravated by violent occipital headaches. During our first examination, the audiogram showed a bilateral deafness clearly predominant on the right side, and the BAEPs were altered (Fig. 1).

The CT scan showed a 4 cm right-sided cerebello-pontine acoustic neurinoma, with no sign of hydrocephalus. Although the BAEPs showed an important elongation of the left I-V period, a non-intracanalar lesion was visible on the left side. Given the great volume of the tumor on the right side, air CT SCAN were not done. 3 months after the first surgery on the right neurinoma, CT control showed a grade II neurinoma on the left cerebello-pontine angle, for which the intra-operative BAEPs will be recorded (Fig. 2).

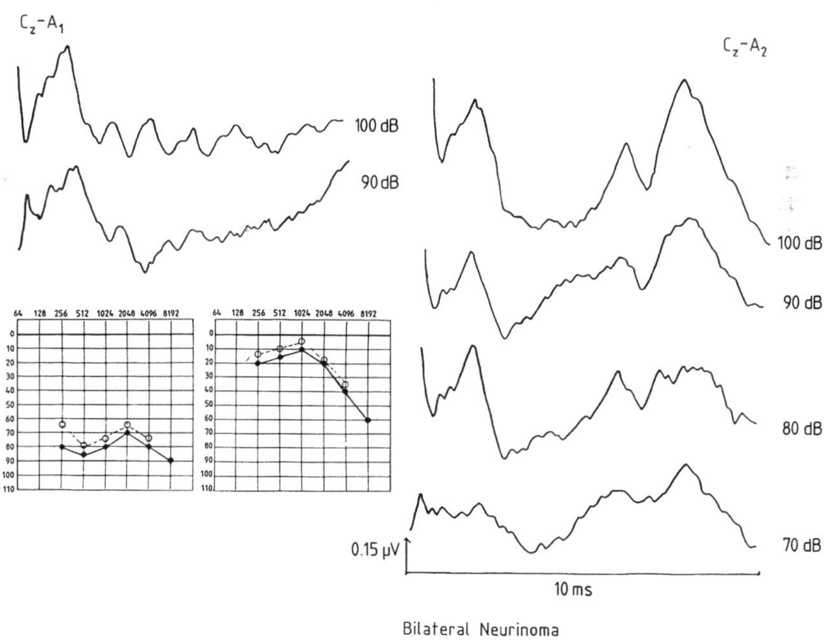

Fig. 1. (Case 1) BAEPs and pure tone audiogram, before the operation. For the rigth ear (Cz-A1) we note a severe deafness and only wave I is visualized, by 100 and 90 dB SPL stimulus. For the left ear (Cz-A2), we note a little deafness with an increased I–V delay (6.2 ms at 100 dB SPL)

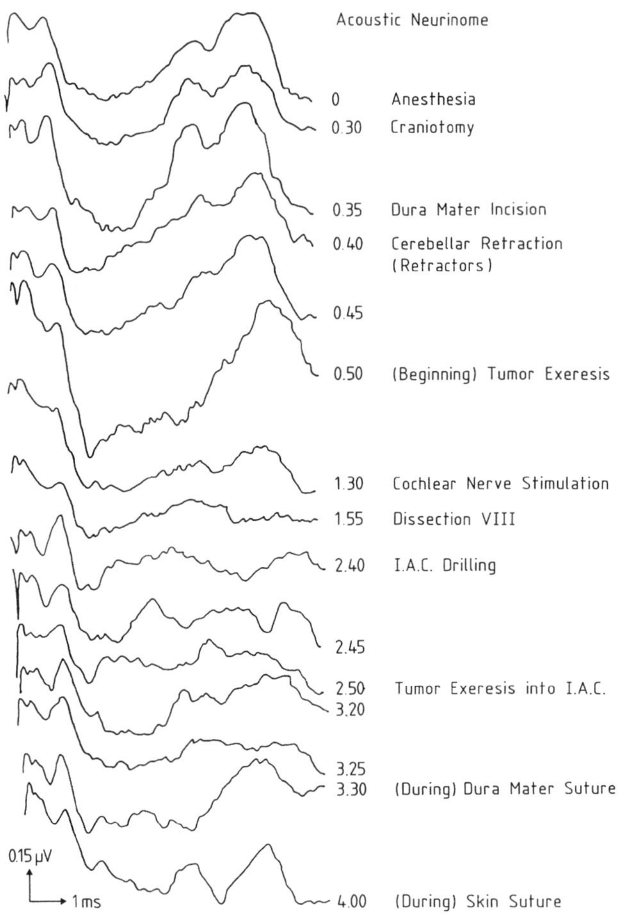

Fig. 2. (Case 1) Per-operative recordings vertex (+), left ear lobe (−) during acoustic neurinoma surgery, in sitting position. The important times which change the pattern of recording are the cerebellar retraction (increased I–V delay) and tumor exeresis, near cochlear nerve. We note ($t = 1.30$) a decreased amplitude and dysnchronized wave V. Wave I, only persists during 90 min. At the end of operation, when retractors are taken out, the wave V re-appears

Intra-Operative BAEPs

It has been noted that during the craniectomy and opening of the dura mater, there is no change in latency or amplitude for wave I or wave V. However, when the cerebellum is retracted, a wave V elongation of about 0.3 ms is noted.

Then, during the tumor exeresis, a group of fibers is isolated corresponding to the cochlear nerve. This period coincides with a latency elongation of 0.3 ms for wave I and 0.6 ms for wave V. As the surgeon looks for the cleavage plane between these fibers and the tumor, the latency increase of wave I continues, the latency of wave V increases also and its amplitude decreases, and the wave V progressively disappears (the recording done with a 20 ms analysis time confirms the disappearance of wave V).

For 90 minutes, only wave I remains reproducible. The rest of the tracing appears desynchronized and wave V is not individualized. This period corresponds to the dissection of the tumor in the angle, to the drilling time for the internal auditory canal and to the exeresis of the tumor at this level.

It is evident that wave V is again recordable at the end of the operation during the closing of the cutaneous planes. Its latency is identical to what it was at the beginning of the operation, and the latency of wave I is sligth delayed (0.5 ms in comparison to the pre-operative recordings).

Comments

Only the recordings ipsi-lateral to the tumor and thus to the operation were made, since for this patient the contro-lateral nerve had been sectioned during a previous operation. It is to be noted that wave V disappears for 90 min while wave I remains throughout the operation. There was a transitory wave V disappearance in 4 other cases of acoustic neurinoma surgery. These disappearances lasted from 10–135 min at the most.

It could be said that in the case of this young woman who presented an acoustic neurinoma with a contro-lateral cophosis, and for whom the issue was the *preservation of hearing* in her only functional ear, the intra-operative recordings were a considerable help in anatomical marking and thus helped preserve her hearing.

An electrocochleography performed 15 days after the operation confirmed the preservation of useful hearing with a 50 dB threshold for unfiltered click stimulation 2 kHz filtred click, and a 60 dB threshold for 1 kHz filtred click stimulation (Fig. 3).

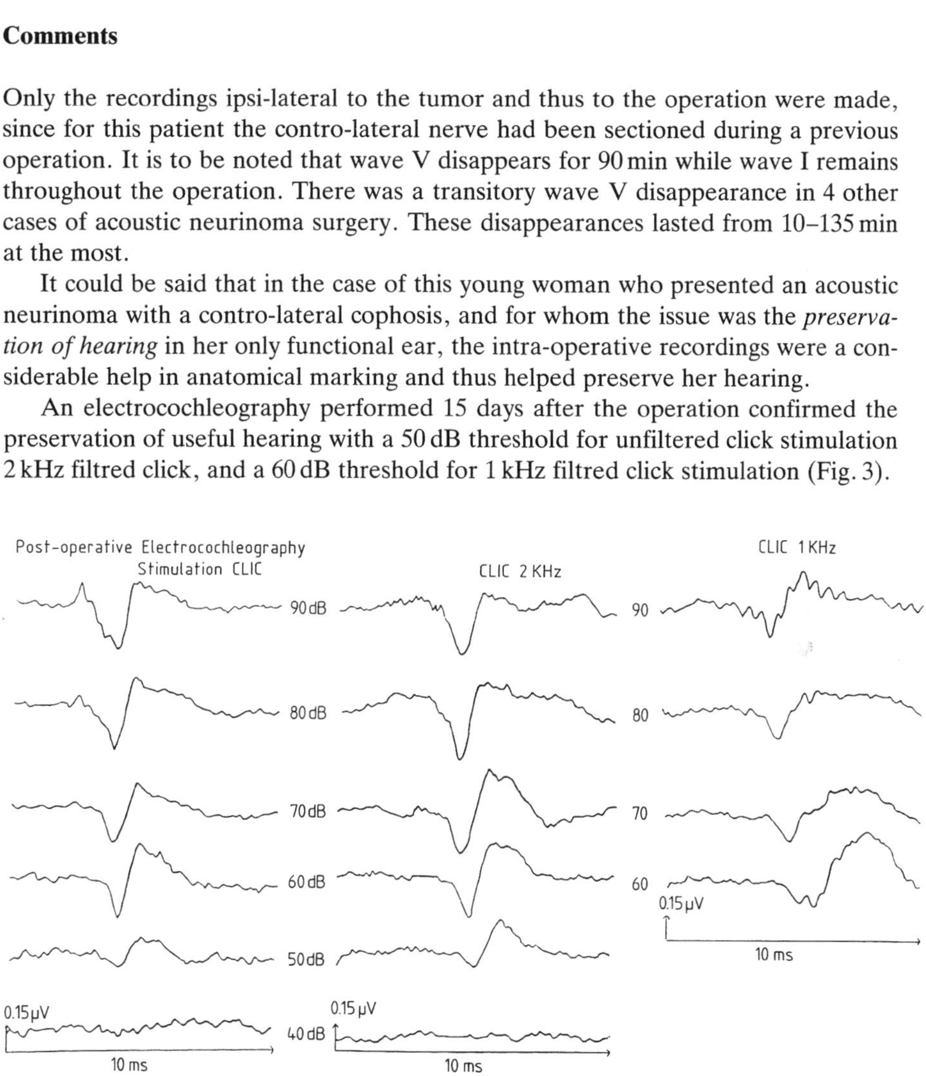

Fig. 3. (Case 1) Left ear, electrocochleography, 15 days after operation. Hearing is preservated and the objective threshold, by a 2 kHz filtred click stimulus as at 50 dB SPL, and by a 1 kHz filtred click is at 60 dB SPL.

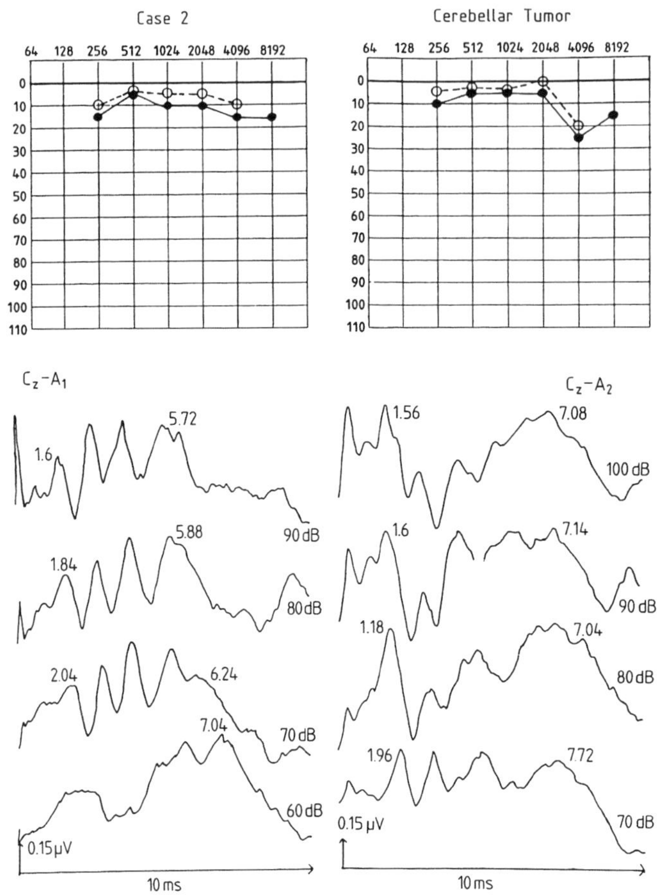

Fig. 4. (Case 2) BAEPs and pure tone audiogram before the operation. For the right ear, the hearing and the recording are normals. For the left ear, we see, 25 dB SPL loss on the 4 kHz frequency. BEAPs shows an increased I–V delay (5.52 ms to 100 dB SPL)

Case No. 2

A 38-year-old man was being treated for positional vertigo associated with headaches. Clinical examination and electronystagmography pointed to a central vestibular syndrome with possible cerebellar lesion.

The audiogram showed a slight hearing loss of 25 dB on the 4 kHz frequency for the left ear.

The BAEPs taken in the outcome of a dizziness syndrome show (Fig. 4) normal patterns for the right ear for the left ear a clear elongation of wave V latency. The I–V delay is 5.52 ms at 100 dB. The I.T. I–V is quite elongated.

The CT Scan shows a cyst formation in the left cerebellar hemisphere, forcing back the fourth ventricle cavity and being responsible for a hydrocephalus.

The operation was performed on the left lateral sub-occipital side and we recorded the intra-operative ipsi and contro-lateral BAEPs.

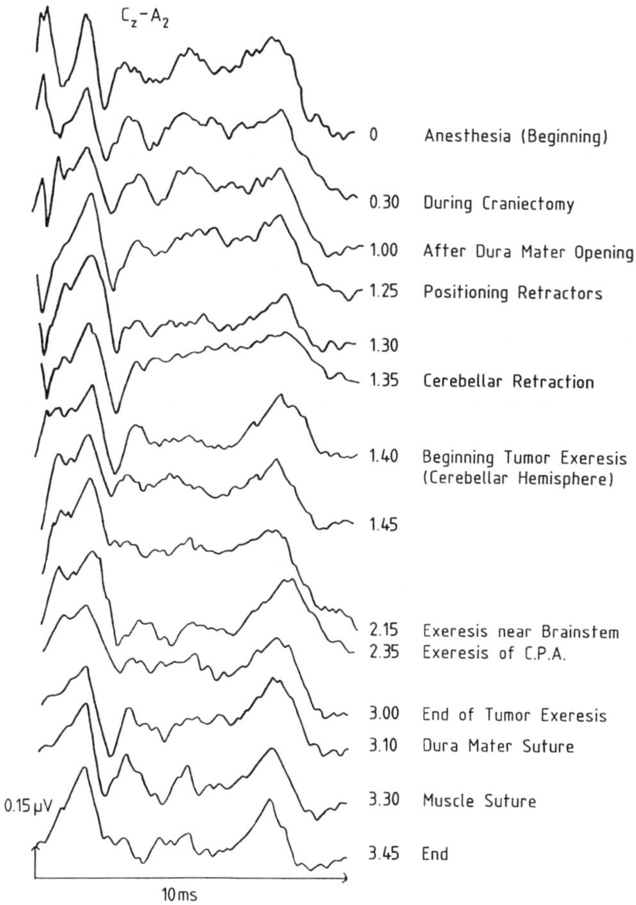

Fig. 5. (Case 2) Per-operative, ipsi-lateral tumor recordings (cerebellar medulloblastoma); vertex (+), left ear lobe (−). The cerebellar retraction, increases the wave V latency (*t:* 1.30) and decreases its amplitude (*t:* 1.35). The tumor exeresis, in its part compressing the brain stem, leads to the I–V delay, second variation

Intra-Operative BAEPs

The ipsi-lateral recordings of the tumor (taken from the vertex − left lobe) are shown in Fig. 5. It should be noted that during the craniectomy and the opening of the dura, there was no change in the latency time or the amplitude of waves I and V.

But, during the retraction of the cerebellum, there was an elongation of 0.35 ms in wave V. This elongation would worsen to reach 0.58 ms in the 5 min following this step, and it was associated with a decrease in amplitude in the same wave.

When the surgical manoeuver was modified, the latency of wave V diminished steadily in the related recordings.

A second alteration of the tracings was noticed during the exeresis of the part of the tumor that was pressing down on the lateral side of the brain stem; this elongation would persist from the end of surgery until the closing of the dura.

During the closing of the muscular and cutaneous planes, wave V latency was seen to decrease progressively and at the end of the operation, the wave V latency was at its pre-operative level and there was a similar I–V delay.

The contro-lateral recordings (taken from the vertex-right lobe) are shown in Fig. 6.

No latency time modification was noted for waves I, II, III or V particularly during cerebellum retraction as in the preceding recordings.

On the other hand, when the surgeon set up a retractor which entailed a pressure at brain stem level, there was pressure on the right side, and the IV–V complex was altered immediately, and this lasted throughout this part of the tumor exeresis. Waves I, II and III did not change, as Fig. 6 indicates.

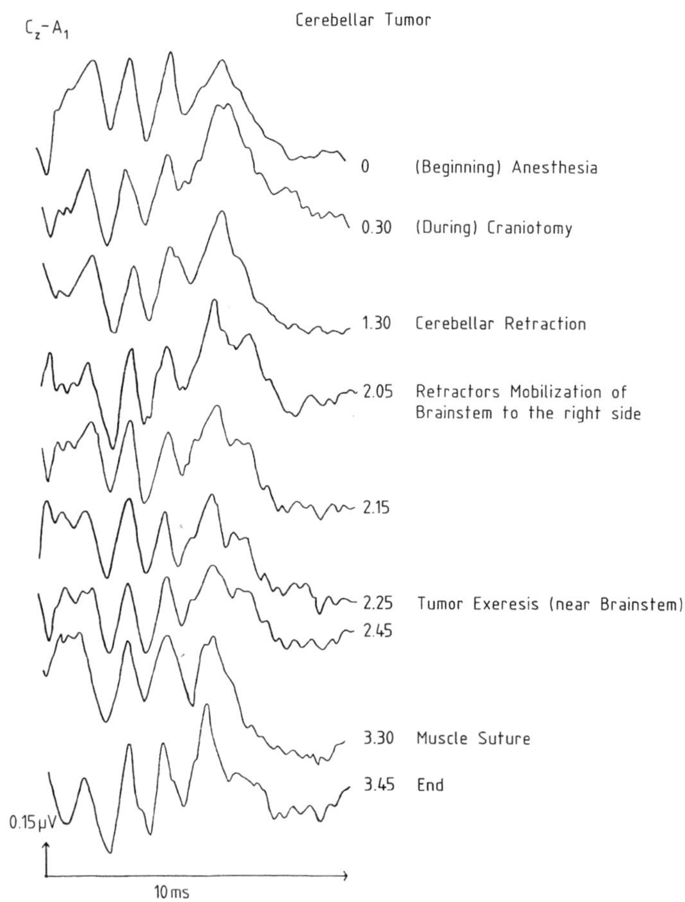

Fig. 6. (Case 2) Per-operative, controlateral tumor recordings; vertex (+), right ear lobe (−). The cerebellar retraction doesn't lead to wave V latency variations as in ipsi-lateral recordings. But, during retractors mobilization, separating the brain stem to the right (side of recording), we note IV–V complex modifications

Comments

In the ipsi-lateral recordings, during the *cerebellum retraction,* a latency elongation of wave V was noted.

Even a light retractor *pressure at brainstem level* brought about a modification of the IV–V complex, rough and immediate in appearance, on the side towards which the brain stem had been pushed slighty (contro-lateral side during exeresis). The ipsi-lateral side modifications came on more slow and were seen some 20 min afterwards.

It should be noted that at the end of surgery, for both right and left recordings, there was a normalization of wave V latency and a return to a I–V delay of the same level as at the beginning of the operation.

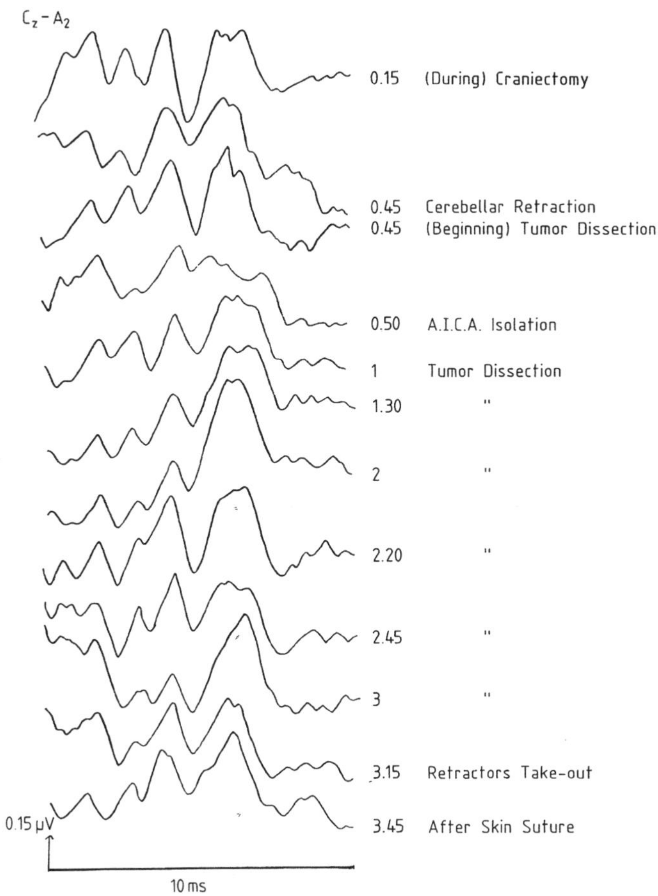

Fig. 7. (Case 3) Pure tone audiogram, BAEPs and electrocochleography before the operation. Right side BAEPs (Cz-A1) show flat recording where any wave is seen. The electrocochleography shows a biphasic N1 peak recorded until 30 dB SPL unfiltered click stimulus. This BAEPs-ECoG discordance traduces a retrocochlear lesion. Left side BAEPs (Cz-A2) see an increased I–V delay

Case No. 3

A 60-year-old woman had been hospitalized for violent headaches which developed in the context of an old deafness of the right ear associated with a vestibular syndrome. The audiogram showed a right deafness predominating in the high frequencies. The electronystagmography revealed an areflexia of the right vestibule.

The BAEPs (Fig. 7) indicate, for the right side recordings, an absence of wave V reproductible at different stimulations carried out at an intensity of 100 dB.

In the left side recordings, an elongation of delay I–V is apparent, which was 4.64 ms at 100 dB.

A right ear electrocochleography was performed which showed a N1 bifid wave of normal latency and amplitude. An electrocochleographic threshold of 40 dB was reached. The discordance of the BAEPs desynchronized tracings as well as the tracings of the response to electrocochleography and the audiogram, indicated a right retro-cochlear lesion which the CT scan showed as a voluminous grade IV neurinoma bearing down on the brain stem. In this case, the right ipsi-lateral tumor recordings could not be viewed because of a lack of pre-operative tracings. Only the controlateral side (where the brain stem compression still entails an elongation of the latency time) was viewed for surveillance.

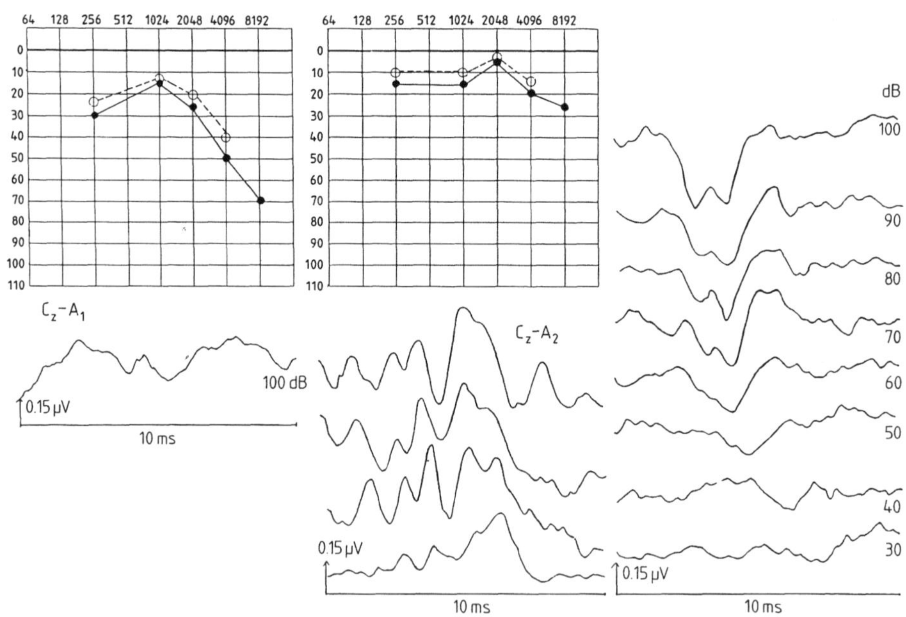

Fig. 8. (Case 3) Per-operative controlateral tumor, recordings (acoustic neurinoma). We note a quick wave V latency lengthening, at the moment of antero-inferior cerebellar artery manipulation. The wave V latency will need an hour to found again its initial value

Intra-Operative BAEPs

On the first recordings (Fig. 8) carried out at the time of the anesthetic induction, the I–V delay was 4.67 ms. During craniectomy, the opening of the dura mater, the cerebellum retraction and the beginning of the tumor dissection, particularly in its lower part near to the mixed nerves, no tracing modification was noted.

On the other hand, during the isolation of the antero-inferior cerebellar artery, a rough elongation of 0.5 ms was brought about in wave V. This elongation was accompanied by an amplitude decrease and lasted for 15 min. The wave V latency then progressively shortened down to the level it had shown at the beginning of surgery, an hour after isolation.

After this episode, no further important variation was noted in wave V. At the end of the operation, it was noticed that as the cerebellum retractors were removed, wave V latency decreases again. Thus following surgery, the I–V delay was lower than 0.2 ms in comparison with the tracings made before surgery.

Comments

In this case, two elements appear significant:

- the manipulation of the *artero-inferior cerebellar artery* led to a sudden elongation of wave V latency of 0.5 ms, which lasted 15 min. This occurred in 4 other cases as well.
- at the end of surgery, the *pressure* of the tumor on the brain stem was removed and wave V latency was clearly decreased in comparison to pre-operative tracings.

Discussion

Our results confirm the recent articles (Grundy et al. 1982, Friedman et al. 1985; Levine et al. 1984; Ojemann et al. 1984; Raudzens and Shetter 1982) which show the intra-operative BAEPs advantages.

If the 26 recordings are globally analyzed, it seems possible to show obvious variations of tracings which are significantly modified during different operative moments.

Chronologically, we noted that:

1. During the *craniectomy,* wave I was stretched (1 case) or disappeared (5 cases). This mechanism was accompanied by an elongation of wave V's latency and the I–V delay stayed normal. Rapidly, wave I reappeared some 5–30 min later.

These wave I modifications seem related to a transitory decrease in hearing. Could an alteration of the cochlear function, with a loss of high frequencies, be suspected from the importance of vibration by the sound traumatism?

2. During the *opening of the dura mater,* in 50% of the recordings during surgery for expansive lesions, a decrease of wave V latency and of the I–V delay was found again.

We have considered as significant variations greater than or equal to 0.25 ms. It must be noted that these modifications were found again in the same frequency in the ipsi- and contro-lateral tracings. The conduction time improvements correspond to a time of maximal decompression after the opening of the cranium and the dura mater. It is to be noted that this improvement of conduction time seems very quickly, in a few minutes only.

3. *The cerebellar retraction* is one of the more critical times for the intra-operative nervous conduction recordings. (Raudzens and Shetter 1982; Grundy et al. 1982). In 80% of the ipsi-lateral recordings at the lesion and in 35% of the contro-lateral recordings, a wave V latency elongation and a I–V delay were noted.

This elongation is from 0.2 ms–0.6 ms depending on the tracings. In some cases, it is immediately stable and remains so. In other cases, it increases progressively forcing the surgeon to replace the retractors while looking for a site less traumatized from an electrophysiological perspective.

4. Let us view the modifications corresponding to *tumor dissection times* in cases of *neurinoma* or meningioma of cerebello-pontine angle (15 cases). It is to be noted that:

– Latency elongations and wave V amplitude decreases are brought about in all cases. However, as in case 1, the disappearance of wave V with persistance of wave I (and in certain cases, without wave I, when it was not present at the beginning of surgery), occurred in 5 cases. This does not necessarily indicate a definitive loss of hearing. It was seen that this disappearance was of a duration of 10 ms–135 ms. However, hearing was conserved in all cases where wave V was again synchronized and registered at the end of surgery (3 cases).

– It was also noted during tumor dissection in the angle that even minimal mobilization of the antero-inferior cerebellar artery created tracing alterations (5 cases). It is either an amplitude decrease, with wave V latency elongation, which is immediate and important (0.3 ms–0.5 ms) or a desynchronization of IV–V complex which becomes difficult to individualize.

5. We have compared these modifications with those obtained during surgery for *3 aneurisms,* of which one was in the vertebral artery and two in the postero-inferior cerebellar artery.

It can be noted that the postero-inferior cerebellar artery dissection doesn't involve important modifications of tracings. More often, it would be a replaced or mobilized retractor up against the cerebellum or the brain stem that would bring about elongations in wave V latency or I–V delay rather than vessel dissection or putting a clip in place.

6. The 3 cases of *microvascular decompression,* compound 2 Vth cranial nerve decompression for facial neuroglia and a VIIth cranial nerve decompression for facial hemispasm.

The point to note seems to be that the different modifications brought about during surgery (cerebellum retraction, vascular loop dissection, muscle interposition) in all cases *regressed very rapidly at the end of surgery* (Fig. 9).

7. We have just analyzed in this discussion the variations observed during the different operational times; if we compare in particular the recordings made at the end of surgery (after closing the cutaneous planes) with those made at the beginning (just after anesthetic induction), it is noted:

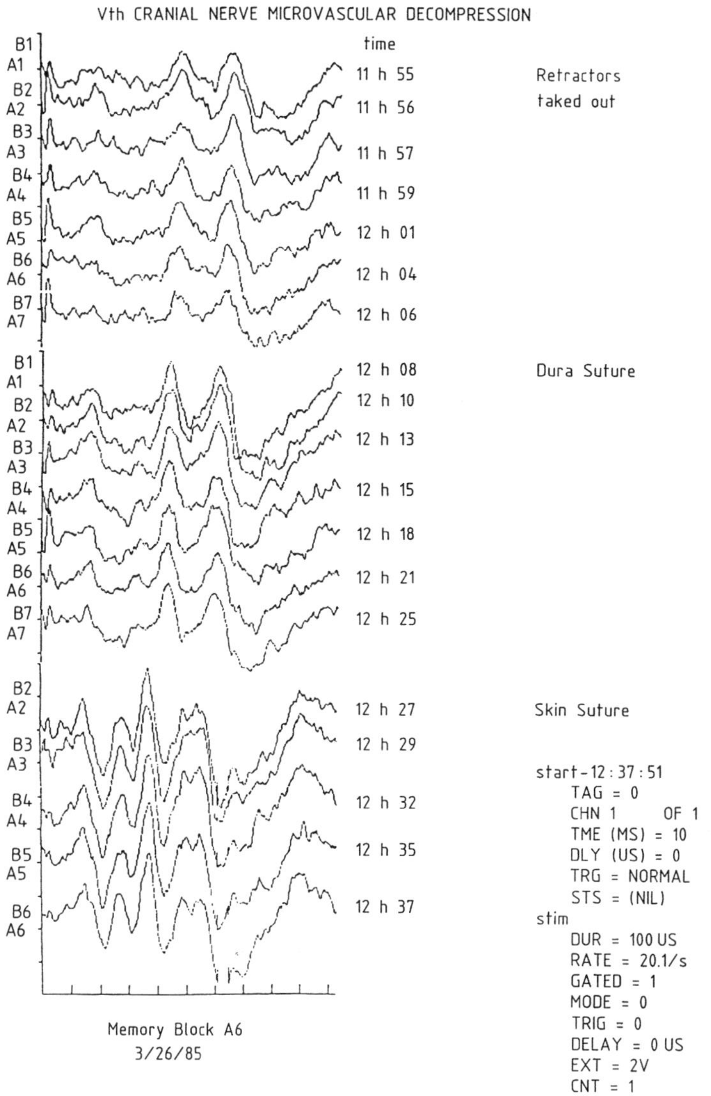

Fig. 9. In this recording it can be noted, that in the last 40 min, at the end of surgery, after the retractors were removed, a rapid progressive decrease in wave V latency and I–V delay occurs, which decreases respectively from 1 ms and from 0.8 ms

a) In all cases, where there was a *reversibility* of the intraoperative alterations of the *I–V delay* with a return to initial values (with a tolerance of 0.2 ms) or even when this parameter was improved, there were no serious post-operational neurological complications.

For example, if we take a lateral sinus wound which develops during surgery, for a superior cerebellar stem glioma, it is to be noted that:

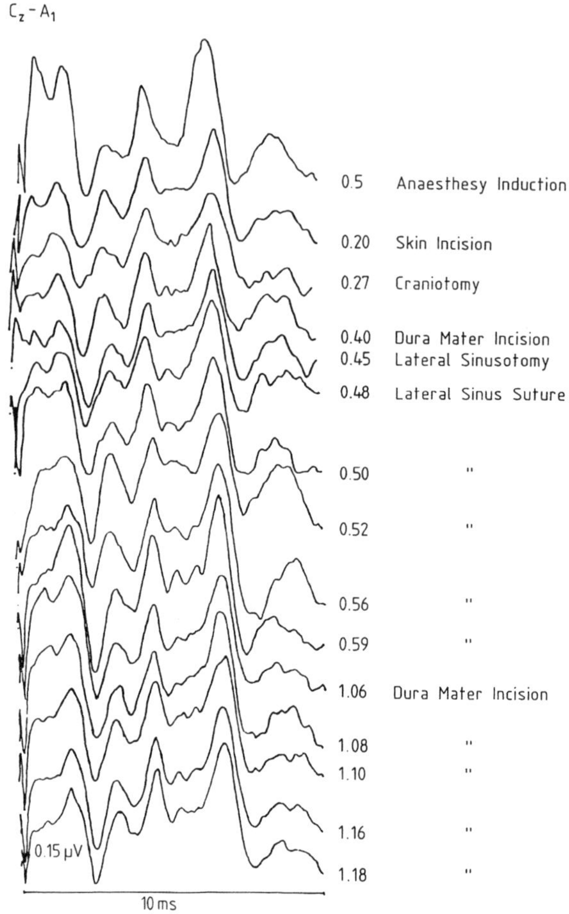

Fig. 10. In the recordings, we can see that a lateral sinusotomy with an important collapse, hasn't involved a stretching or a desynchronization of different waves or I–V delay

– a collapse accompanied by rough drop of arterial tension did not bring about a noteworthy modification of the I–V delay or of synchronization (Fig. 10);

– at the moment of tumor dissection (Fig. 11) at the level of the high portion of brain stem, an elongation of wave V, an amplitude decrease and a desynchronization of the entire tracing occurred. At the end of the surgery, we noted that the recordings have begun a distinct improvement. There were no operative after-effects.

b) On the other hand, in both cases where the I–V delay demained elongated after surgery with no tendancy to decrease, in one case, the patient presented a brain stem infarct and in the other case a transitional deficiency with hemiplegia and problems with consciousness. He recovered in one week and is presently well and the BAEPs are normalized.

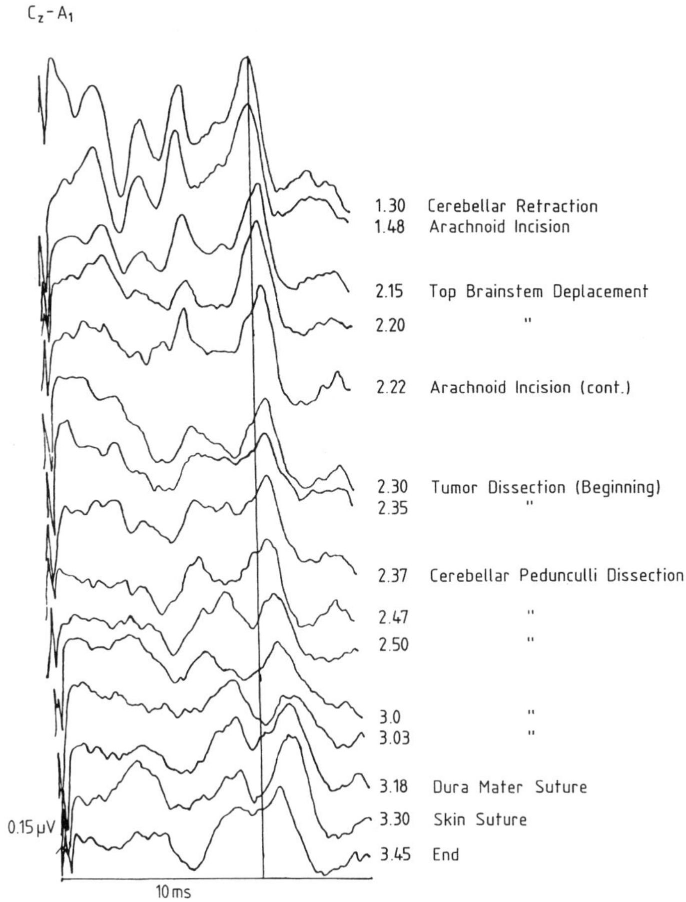

Fig. 11. In these traces recorded after lateral sinus suture, we can see a clear progressive stretching of wave V latency, with a decrease in amplitude. These modifications correspond to the different tumor dissection times. During the last recordings, in the moment of sutures (dura, skin), a decrease tendancy of wave V and I–V delay was noted. 8 days after the surgery, the BAEP's were normalized and no neurologic complications were noted

Thus, it seems possible to say that there exists a certain anatomo-electrophysiological correlation, which, despite a number of limits, allows these recordings to be given a functional prognostic signification.

In conclusion, intra-operative monitoring of the auditory and the brain stem functions, such as it has been utilized up to now, allows early surveillance of the vital functions of the brain stem by ipsi- and contro-lateral recordings and surveillance of the auditory function only in cases where preoperative tracings were present, which implies a minimal residual hearing of high frequencies.

References

Friedman WA, Kaplan BJ, Gravenstein D, Rhoton AL (1985) Intra-operative brain stem auditory evoked potentials during posterior fossa microvascular decompression. J Neurosurg 62:552–557

Grundy BL (1982) Monitoring of sensory evoked potentials during neurosurgical operations: methods and applications. Neurosurgery 11:556–575

Grundy BL, Jannetta P, Phyllis T, Procopio BA, Lina A, Boston R, Doyle E (1982) Intra-operative monitoring of brain stem auditory evoked potentials. J Neurosurg 57:674–681

Hardy RW, Kinney S, Lueders H, Lesser RP (1982) Preservation of cochlear nerve function with the aid of brain stem auditory evoked potentials. Neurosurgery 11:16–19

Hashimoto I, Ishiyama Y, Yoshimoto T, Nemoto S (1981) Brain stem auditory evoked potentials recorded directly from human brain stem and thalamus. Brain 104:841–859

Levine RA, Ojemann RG, Montgomery WW, McGaffigan P (1984) Monitoring auditory evoked potentials during acoustic neuroma surgery. Ann Otol Rhinol Laryngol 93:116–123

Moller A, Jannetta P (1983) Interpretation of brain stem auditory evoked potentials: results from intracranial recordings in humans. Scand Audiol 12:125–133

Ojemann RG, Levine RA, Montgomery WW, McGaffigan P (1984) Use of intraoperative auditory evoked potentials to preserve hearing in unilateral acoustic neuroma removal. J Neurosurg 61:938–948

Raudzens P, Shetter AG (1982) Intra-operative monitoring of brain stem auditory evoked potentials. J Neurosurg 57:341–348

Zappulla R, Greenblatt E, Kaye S, Malis LA (1984) A quantitative assessment of the brain stem auditory evoked response during intraoperative monitoring. J Neurosurg 15:186–191

Advances in CT and NMR Stereotaxic Systems and Related Methods of Radio Frequency and Radiation Therapy

E. R. Cosman

Once CT scanning became routine, it was inevitable that it would be united with stereotaxy to produce an exciting new capability in neurosurgery. That has happened in the last 2 years, and we have witnessed a revolution in brain stereotaxy. An indication of this is that for one CT stereotaxic guide with which I am most familiar, the Brown-Roberts-Wells (BRW) CT Stereotaxic System, there have been 120 units put into operation in the United States alone between 1982 and 1985. Most of these are being used by neurosurgeons who never practiced stereotaxy before. I will summarize here what is currently being done with this system, and some of its potential uses for the immediate future.

First, let me describe briefly how the BRW instrument works. Fig. 1 shows its modular elements (Brown et al. 1980; Heilbrun et al. 1983). A head ring fastens to the skull by sharpened screws on carbon fiber posts which swivel to allow any orientation. Onto the head ring is placed the removable localizer ring (Fig. 1a) with its "N-type" carbon rod structure. A CT scan through the localizer at any orientation provides rod index image points to determine fully the CT plane and any target in the head within it. Entry of CT data into a hand-held preprogrammed computer enables AP, lateral, and vertical coordinates of the target as well as arc angle settings to be calculated. Fig. 1b shows the Arc System attached to the head ring. The calculated target and arc angles are confirmed prior to probe insertion by the phantom base

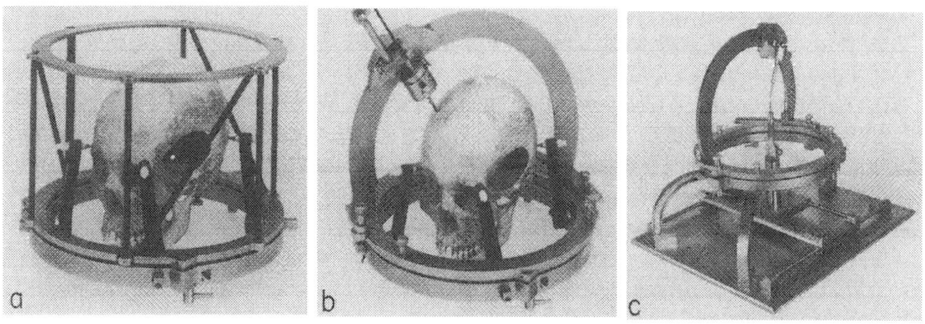

Fig. 1a–c. Some of the basic modular elements of the BRW Image Guided Stereotaxic System. **a** head ring attached to the skull by 4 posts with sharpened screws, and the CT localizer ring attached to the head ring to provide fiducial image points during the CT scanning phase of the procedure. **b** Arc System attached to the head ring to provide guidance during the surgical phase of the procedure. **c** Arc System attached to the Phantom Base which enables confirmation of both the calculated target position and the Arc System angle settings prior to probe insertion into the brain

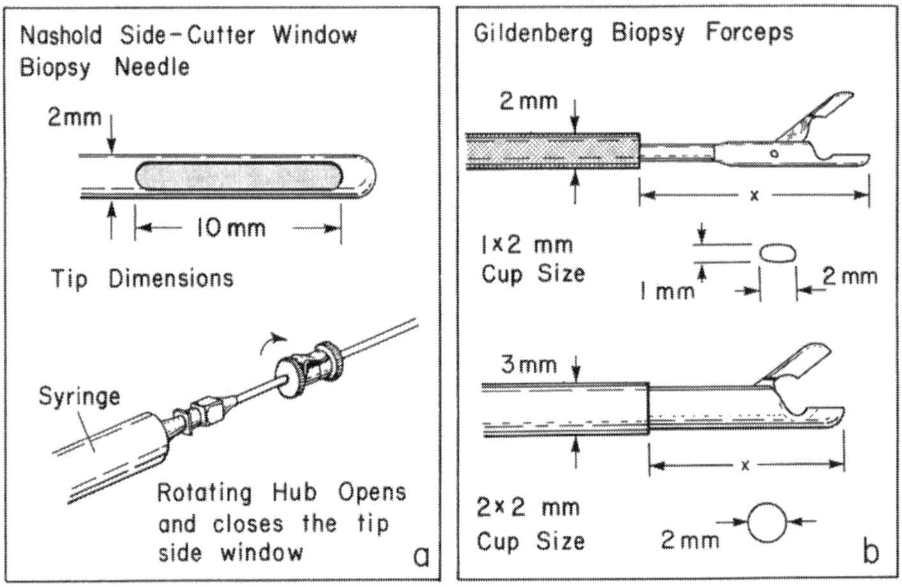

Fig. 2a, b. 2 types of biopsy instruments. **a** The side-cutting window type biopsy needles have been specified by Dr. B. Nashold (of the dimensions shown) and by Dr. R. Sedan. **b** The biopsy forceps. The tips shown are of the Gildenberg styles (from Radionics, Inc.), and there are other styles available (such as from Fischer GmbH)

which also accepts the arc (Fig. 1c). The arc enables approaches from any angle without obstruction, and overall accuracy is typically less than 1 mm.

Among currently active applications of this device are tumor biopsy, radionuclide implantation, hematoma evacuation, and abscess drainage. Many new ancillary instruments have emerged for these procedures, and I will discuss some of them here based on a survey of over one hundred practitioners. Fig. 2 illustrates commonly used biopsy instruments. The side-cutting window type biopsy needle (Fig. 2a) is most popular and gives excellent specimens of about $1 \times 1 \times 10$ mm size. It is ideal for soft tissue tumors. Where hard, granular tumors are encountered, the biopsy forceps (Fig. 2b) are required. Though the 1×2 mm cup is discrete, its samples are marginally small, and the 2×2 mm cup forceps is often preferred. Other systems, such as sharpened front-cutting aspiration needles or coil biopsy devices are used successfully, but are not as well thought of as the designs of Fig. 2.

Stereotaxic hematoma evacuation has been performed at several centers, though it has not yet gained wide acceptance. The Higgins-Nashold instrument (Higgins et al. 1982) of Fig. 3 is an improvement of the Backlund Screw design. Its spiral cutter, which rotates by turning a knob on the hub, faces the side window at its tip enabling bits of congealed hematoma to be sheared off. Internal irrigation and aspiration channels gently flush the cutter's flutes and draw the debris up and out of the shaft. This feature is essential for proper action.

A variety of stereotaxic drainage catheters for the BRW have been designed by Drs. G. Broggi (Milano), R. Martuza (Boston), and M. Apuzzo (Los Angeles), and

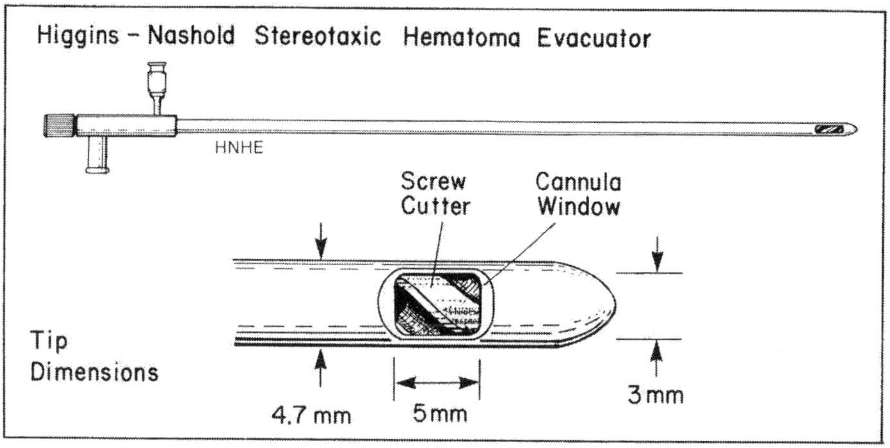

Fig. 3. The hematoma evacuator. The design shown is specified by F. Higgins and B. Nashold and has a side-window and screw-type cutter with both irrigation and aspiration channels internal to the cannula. The earlier Backlund design was similar, but with only an aspiration channel

these have been very effective in entering and draining small, difficult to reach cysts, abscesses, and ventricular volumes located by CT scans.

CT stereotaxy has produced a resurgence of interstitial irradiation (brachytherapy) for treatment of brain tumors. Radionuclide implantation was pioneered by Mundinger and by Talairach using stereotaxic angiography and ventriculography. Now, with easy CT preoperative knowledge of tumor size, many neurosurgical groups are re-addressing stereotaxic brachytherapy. Iodine (^{125}I), Iridium (^{192}Ir), and Gold (^{198}Au) are in common use. I will describe here the mechanics of 2 popular methods of radionuclide implantation. Gutin et al. (1981) of University of California (San Francisco) use high activity ^{125}I sources and a double catheter after-loading system (Fig. 4a). The outer silicone catheter is stereotaxically guided in the BRW arc to the target (Fig. 4b). A fixation sleeve, which is bonded to the skull and to the catheter by biological adhesive, secures the catheter at its proper depth. The inner teflon catheter, preloaded with ^{125}I seeds by the radiologic team, is inserted into the outer catheter, frozen in place by the biologic glue and left for the dose period. Apuzzo and Sabshin (1983) University of Southern California (Los Angeles) use a single catheter technique for ^{192}Ir brachytherapy (Fig. 4c). A silicone catheter is stereotaxically placed and immobilized by a fixation sleeve which is sutured to the scalp and glued to the catheter by Aron alpha-acrylic glue. An ^{192}Ir ribbon, supplied commercially, is afterloaded into the catheter, and the entire assembly secured by a large vessel clamp around the neck of the fixation sleeve. In both techniques careful design of catheter size and durometer and matching guide tubes are essential for accurate effective usage.

The need for CT stereotaxic placement of parallel arrays of probes, such as radionuclide catheters, epilepsy electrodes, hyperthermia electrodes, etc., has led to important ancillary developments. A few of these for the BRW system are noteworthy. Fig. 5a shows a modified offset arc and grid system developed by D. Wright and coworkers at the NIH (Bethesda) for radionuclide implantations. Replaceable grids

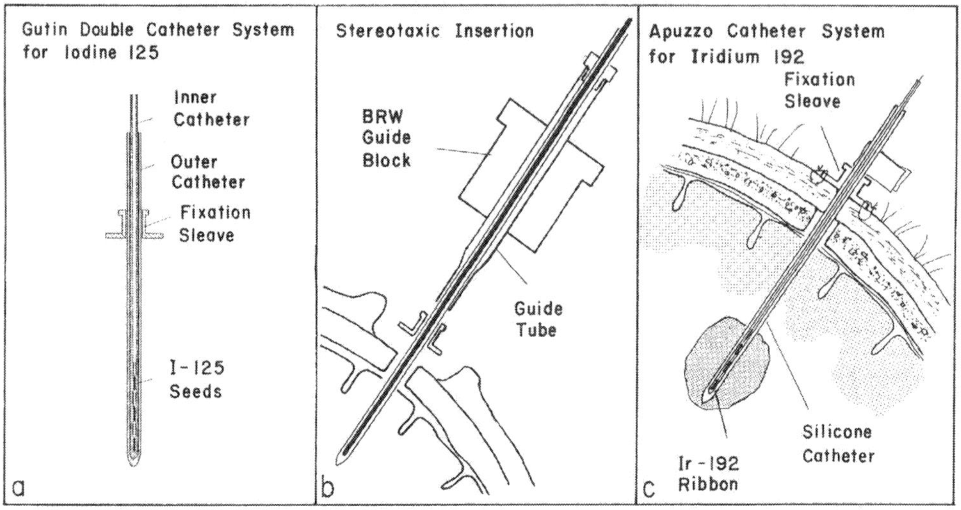

Fig. 4a–c. Radionuclide catheters. **a** The double after-loading design of Dr. P. Gutin has an outer silicone catheter and inner teflon afterloading catheter which is preloaded with Iodine 125 seeds. **b** Implantation with the BRW Guide. The fixation sleeve is secured to the catheter with biological adhesive and sutured to the scalp for stabilization. **c** The Iridium 192 catheter of Dr. M. Apuzzo

Fig. 5a–c. Modular accessories to the BRW System. **a** The NIHA Arc with a grid-hole platform for multiple parallel catheter or electrode placement. **b** The Talairach-type guide plate with multiple hole grids. **c** The repeat fixation head posts having metric scales on the drives and calibrated carbon rods to set into skull holes

with any desired interhole distance enable a central stereotaxic ray to be determined, and then a parallel lattice of probes to be passed without arc resetting. This same NIHA Arc, with its ± 4 cm clearance about the central ray, is also being used as a wide-bore platform for computer-controlled laser surgery on tumors by M. P. Heilbrun et al. (Salt Lake City). Fig. 5b shows a Talairach grid guide module for the BRW. It mounts directly onto the BRW head ring and enables X-ray radiography for epilepsy electrode implants to be done just as it is for the Talairach Frame. X-ray reticules and cassette holders mount to this Talairach-type base for doing 3-D X-ray

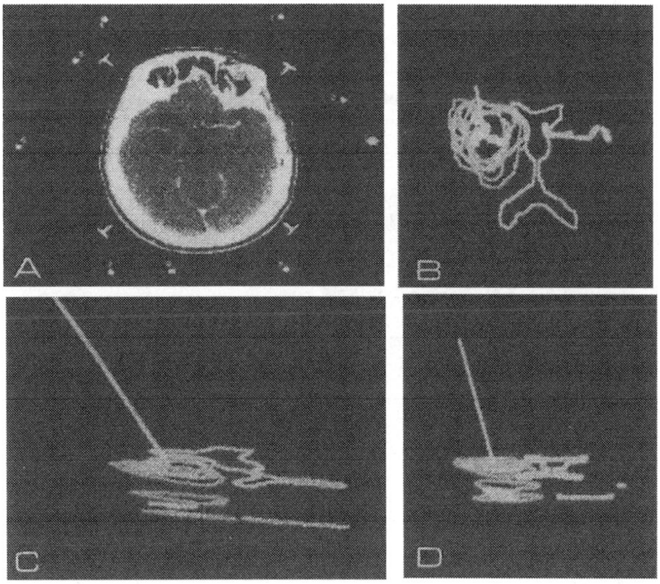

Fig. 6. Volume reconstruction of tumor and selected physiologic structures using computer graphics as developed by Dr. P. Heilbrun

imaging and stereo-angiography, all referenced and comparable to CT determined coordinates on the same BRW head ring. This is a good example of the power of the BRW modular component concept. Finally, Fig. 5c shows a repeat-fixation means for the BRW head ring enabling a patient to be placed in the ring in precisely the same position at widely different times for repeat implantation or comparative imaging. Metered head post drives and calibrated carbon skull pins are necessary.

Planning stereotaxic surgery from multiple CT slice images, for example to optimize placement of radionuclide sources, demands more sophisticated data handling and volume reconstructions. Dr. P. Heilbrun and co-workers, University of Utah (Salt Lake City), are developing software for a minicomputer that does this economically and offline, even in the operating room. Scan data are transferred on floppy disk from standard CT or MRI scanners to the minicomputer and brought up on its CRT screen (Fig. 6a). Localizer points are automatically read to put data in BRW frame coordinates, and a "joy-stick" control enables tumor, ventricle, large vessels, etc. to be outlined in various colors. Now a probe tract or tracts are planned on the CRT screen (Fig. 6b, c, d) while viewing it relative to these structures from any angle. Depths and angle settings for the BRW arc are immediately output for surgery.

Stereotaxy based on MRI (nuclear magnetic resonance) scanning is beginning and shows great potential. Fig. 7a shows an MRI localizer for the BRW of Cosman-Wells design, which uniquely enables targets from axial, sagittal, or coronal scans to be determined (patent applied for). It has N-type structures on top, back, front, and sides made of channels in which various para-magnetic solutions may be filled through inlet ports. Thus, an image plane may be reconstructed, no matter what its orientation to the head ring. Parasagittal scans, for instance, should eliminate the

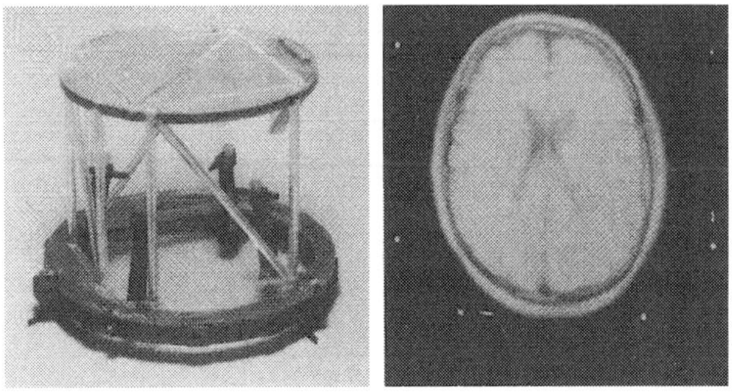

Fig. 7. The MRI magnetic resonance imaging localizer used with the BRW head ring. A cubical array of rods and diagonals enables axial, sagittal, or coronal slice target determinations at any slice angle (Cosman-Wells design)

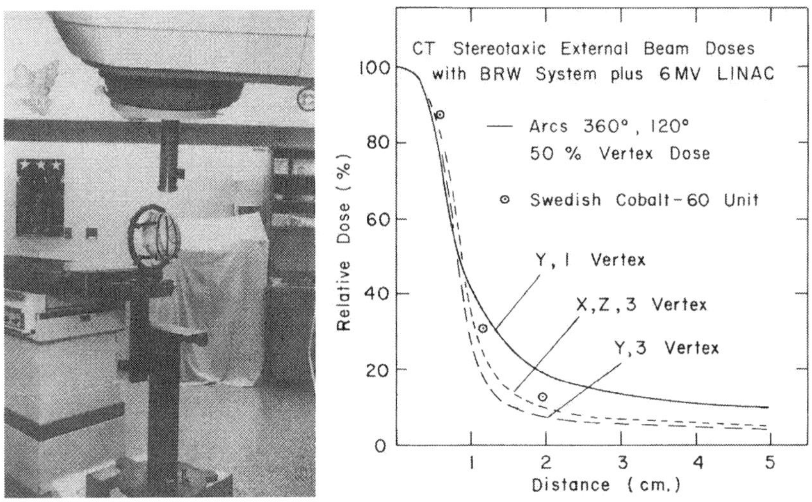

Fig. 8a, b. Focal external beam irradiation based on CT, MRI, or angiographic imaging with the BRW system. **a** The BRW floor stand moves the target to the isocenter of gantry and table rotation of an electron LINAC gamma ray source. **b** Comparison of dose profile away from a target using the LINAC produced external gamma beam source to that from the Swedish "Gamma Knife" source which utilizes multiple fixed Cobalt 60 sources

older lateral ventriculograms for functional surgery. Fig. 7b shows an axial MRI localizer scan done with this device for surgery by Dr. D. Thomas of Queen's Square Hospital (London).

External beam irradiation is another technique being transformed by the union of CT and stereotaxy. Many radiotherapy centers already have electron LINACs which can produce rotatable, variable energy highly collimated photon beams. Together with the BRW CT stereotaxic system, a small field, high dose radiation of any volume seen on a CT or MRI scan, can be delivered. Fig. 8a shows the arrangement now being set up by Drs. K. Winston, W. Lutz, and co-workers at Children's and Brigham and Women's Hospitals (Boston). The BRW Floor Stand (comes standard

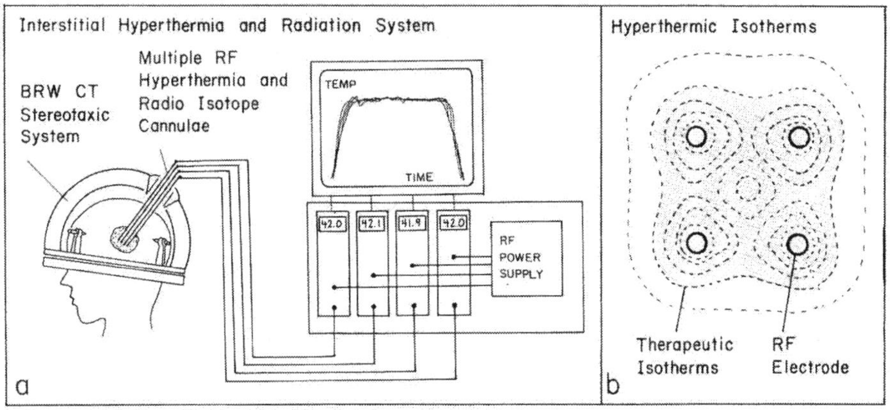

Fig. 9a, b. Interstitial hyperthermia and brachytherapy. **a** Schematic of a computer-controlled multiple radiofrequency electrode system by E. Cosman. **b** Typical isotherm lines produced from an array of four rf-electrodes

with the BRW) is mounted on a platform which rotates about a vertical axis, as does the patient table. A 6MV Varian LINAC gantry varies the declination angle of a collimated photon beam from above. The coordinates of a tumor are identified with the BRW from a CT scan. The BRW Floor Stand enables precise x,y,z translation of those coordinates from 0,0,0, so that the tumor can be brought to the exact isocenter of the platform and gantry rotations. Winston and Lutz have devised elegant methods of checking this using the BRW Phantom Base. Now, by a series of platform and gantry rotations, the photon beam is spread out over the hemisphere of the patient's head, but delivers a high summed dose at the focal isocenter. Fig. 8b shows test results on phantoms. Doses of 5000 to 10,000 Rads can be delivered to 1 cc volumes. With proper collimation to reduce penumbra, the dose levels at the target edge drop from 80%–20% in 2 mm.

I will mention finally two emerging examples of CT stereotaxy of great interest: interstitial hyperthermia and CT body stereotaxy. Fig. 9 illustrates schematically a multi-rf-electrode hyperthermia system which I have been building for brain or body tumor treatment. A 300 watt rf power, supply drives the electrodes, and each is thermally monitored under computer control to regulate the thermal uniformity of the heated volume. The therapeutic isotherms are similar to dose profiles in brachytherapy, and the same stereotaxic methods and planning apply as discussed above. Though much must be learned about safety and effectiveness of such a technique, the experience gained from the use of functional rf-heat-lesion of the last three decades provides a valuable guide at the outset. We are developing catheters which deliver in combination rf-heat, radionuclide sources, chemotherapeutic agents, and UV light for multi-modal tumor therapy.

The same positive impact that CT stereotaxy has had on surgery of the brain is about to happen for the rest of the body. Although CT body stereotaxis is primarily the domain of the radiologist, certain areas such as the spine, neck and disks are still for neurosurgeons, and thus, it is relevant to this article. Because the body has no rigid clamping structures like the skull is to head, body stereotaxis poses new prob-

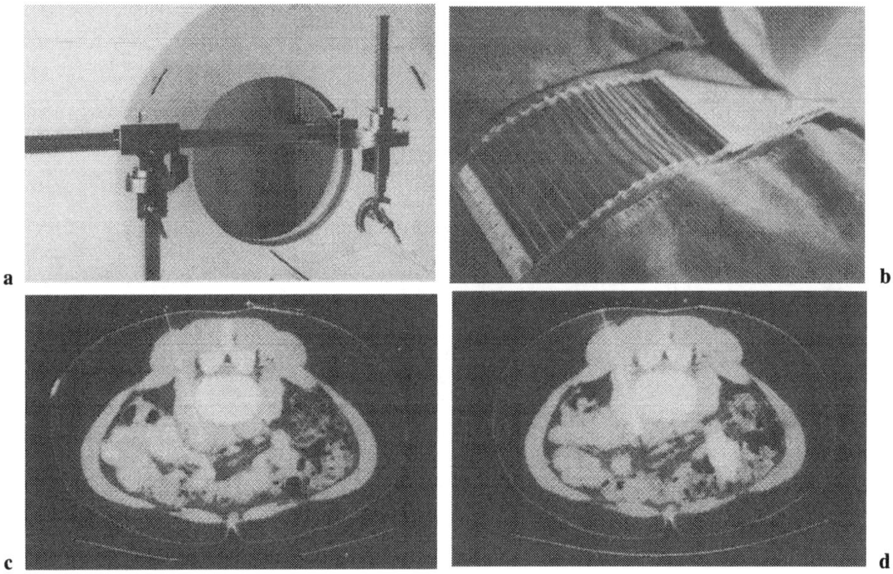

Fig. 10a–d. Image guided stereotaxy in the body. **a** A guidance system stereotaxic probe placement in the body by G. Onik, E. Cosman, and T. Wells. **b** A triangle-plus-ladder type skin localizer by G. Onik and E. Cosman. **c** Planned probe path on the CT scanner. **d** Actual probe path using the stereotaxic guide

lems. A solution which works very successfully is shown in Fig. 10. This is the Onik-Cosman-Wells (OCW) Stereotaxic System (Radionics, Inc., patents applied for), and it comprises a guidance system mounted on a floor stand which is aligned to the CT scanner. Linear x,y,z travels and azimuth and declination angle arcs (Fig. 10a) enable universal direction of a probe holder. A skin localizer is placed on the patient comprising a triangle element and a "ladder" of lateral rods (Fig. 10b). The width of the triangle image from the CT scan determines the axial "level" of the CT plane, and the rods give a multiplicity of choices for a skin entry point for a probe. When a target image point is chosen, the probe path is set up on the scanner (Fig. 10c) and the arc angles read out. The x-y-z travels move the probe tip to the desired entry point, and the insertion can be made. Oblique, or out-of-plane, approaches are also easily set up. Fig. 10d shows the actual probe tract set up in Fig. 10c done in a recent biopsy by Onik and Goldberg at UCSF (San Francisco), in this case hitting a 1.5 cm abnormal lymphonode in the retro-peritoneum. A recent series of over 50 stereotaxic biopsies done at UCSF and the N. E. Deaconess Hospital (Boston) with the OCW system show it to be spectacularly more accurate and faster than freehand technique.

The examples cited here illustrate that we are in a new era of CT stereotaxy and surely its frontier will continue to advance rapidly in the years to come.

References

Apuzzo ML, Sabshin JK (1983) CT guidance stereotaxis in the management of intracranial mass lesions. Neurosurg 12:277–284

Brown RA, Roberts TS, Osborn AG (1980) Stereotaxic frame and software and computer software for CT-directed neurosurgical localizer in invest. Radiology 15:308–312

Gutin PH, Phillips TL, Hosobuchi Y (1981) Permanent and removable sources for the brachytherapy of brain tumors. Int J Radiat Oncol Biol Phys 7:1371–1381

Heilbrun MP, Roberts TS, Apuzzo ML, Wells TH, Sabshin JK (1983) Primary experience with Brown-Roberts-Wells (BRW) CT stereotaxic guidance system. J Neurosurg 59:217–222

Higgins AC, Nashold BS, Cosman ER (1982) Stereotaxic evacuator of primary intracerebral hematomas. Appl Neurophysiol 45:438–442

Onik GM, Cosman ER, Wells TH, Moss AA, Goldberg HI, Costello P (1985) CT body stereotaxic instrument for percutaneous biopsy and other interventive procedures. Invest Radiol (accepted for publication)

Advances in Stereotactic Instruments

G. Broggi

The previous chapter described and discussed the most recently built stereotactic instrument (BRW). It seems to the editors, however, that there is still space for a brief review of the technological improvement of the other instruments most used throughout the world.

For the sake of historical accuracy we must mention that between 1947 and 1967, 18 different devices were developed (Fox 1969), and many others have followed in recent years; but certainly the four most widely-used systems are the Talairach (1949), the Riechert-Mundinger (1955), the Leksell (1949), and the Todd-Wells. All these instruments were built when neuroradiological examination methods consisted of angiography and ventriculography and all standard stereotactic procedures were performed with the aid of orthogonal X-ray equipment.

The development of the computerized scanning and imaging techniques, such as computed tomography (CT), positron emission tomography, and, recently, nuclear magnetic resonance (NMR; also called magnetic resonance imaging), induced surgeons used to stereotactic operations to ask for modification of the systems: today, indeed, all these instruments may be provided with accessories that make them compatible with those new neuroradiological machines (Bosch 1986).

The Talairach System

A special place in this short review has to be granted to the stereotactic system of Talairach. First of all, it is based on outstandingly precise anatomical work. Moreover, the anatomical plans are presented in such an order that every anatomical structure and its relationship with neighbouring areas may be easily recognized.

In his pioneer work (Talairach et al. 1957) Talairach clearly explains how some cerebral structures, like the horn of Ammon, may be "directly" recognized while others, like the amygdaloid nucleus, may be "indirectly" localized; the same methodology was then applied to the whole telencephalon (Talairach et al. 1967). In this very original and hitherto unsurpassed atlas, the author studies the statistical distribution of the cortical regions as function of the well-known "bicommissural line," in order to facilitate the move from a theoretical brain to the individual brain which is the object of the stereotactic surgery. Indeed, the utilization of the "quadrillage proportionnel" based on the individual bicommissural line and on the bone landmarks makes it possible to recognize where the different cortical areas are located and also to go back from the individual anatomical organization to the general statistical schematization, based on the three perpendicular planes in which the brains composing the atlas were sectioned.

The Talairach system is formed by a headframe of rectangular shape that is fixed to the skull and to the operating table. Four small holes are drilled in the skull, allowing precise fixation with pins. This makes the system different from others, because it may be replaced in exactly the same position at various intervals of time during following or reoperation. The X-ray equipment requires a tube film distance of 5 m, giving an enlargement of 2.5% for each target. Double grids are mounted on the headframe, allowing the electrode or biopsy instrument to be inserted through a small trephine opening into the skull orthogonal or parallel to the X-ray axis. With this technique it is possible to place multiple electrodes in the target area. All the instruments are parallel to each other and should have their own trephine holes.

Using this particular approach with this instrument, it is possible to reach almost every anatomical complex in the brain, following the orthogonal plans, knowing not only where a given structure is located, but also what nucleus or system of fibers will be crossed on the way to the desired target. This is definitely a necessity in the short- and long-term placement of electrodes for the stereo EEG recording in the examinations before surgical operation in epilepsy (Bancaud et al. 1965, 1973). Furthermore, simultaneous stereotactic, stereoscopic angiography (Szikla et al. 1977) and ventriculography (Talairach et al. 1957) greatly facilitate access to multiple targets without major risks (Munari 1986) and help in the choice of the most appropriate anatomic and functional approach.

Recently, a modification allowing a polar approach as well has been described; this is useful in the management of deep-seated tumors (Scerrati et al. 1984). Finally, this instrument is the basis for the stereotactic multiple beam irradiation system built by O. Betti, discussed elsewhere in this book.

The Riechert-Mundinger System

Like the Talairach, the Riechert-Mundinger system consists of a circular headframe that leaves the skull convexity totally free, allowing open supratentorial craniotomy. A film distance of 3–4 m is suggested, producing a 3.5% enlargement. The aiming device that guides the introduction of electrodes or biopsy instruments consists of one arc fixed on the transverse interaural axis, allowing placement at any desired angle from 0° to 180°. A phantom device facilitates the calculation of many trajectories passing through the same 10-mm bur hole. With this system radial, polar approach is the rule; the orthogonal approach, although possible, is complex.

This system was made CT-compatible in 1982 (Birg and Mundinger 1982) and NMR-compatible more recently, allowing easier calculation of dosimetry in cases where interstitial radiotherapy is required. The pins of the headframe have been modified, allowing perfect alignment (0°-angle) in the gantry of the CT scanner and further scanning parallel to the ring. Electronic elaborations with midsagittal plane reconstruction also permit stereotactic surgery of "nonvisible" targets; open microsurgery, stereotactically guided is also possible (Jacques et al. 1980).

The Leksell System

This system consists of a cubic headframe fixed to the skull with three of four pins. This headframe is not fixed to the operating table but only to the skull of the patient, which can be placed in any position on the operating table since the X-ray system is coupled to the frame and the target calculation is done with the aid of a localizing diagram that allows a greater X-ray enlargement (20–30%). This feature make this instrument completely different from the first two.

The aiming device, with an instrument carrier, is a semicircular arc fixed on the headframe according to the calculated coordinates of the target, such that the center of the arc corresponds to the target point. The instrument carrier (as for the Riechert-Mundinger apparatus) can be moved along the arc, allowing the surgeon to penetrate to many different targets through the same bur hole. In this system the target is always positioned to the 0-point of the coordinates, due to the possibility of an orthogonal movement of the pivot of the arc.

This system has also been adapted to direct use of CT, the headframe being fixed to the moving table of the scanner. The fixation pins are made of carbon fibers, as in the other instruments, and a pair of plastic walls with a N-shaped wires are fixed to the frame, allowing computerized calculation of the coordinates directly from the CT imaging. An NMR-compatible headframe has also recently become available. A modified new headframe in the shape of an incomplete octogonal ring suitable for the cubic frame attachment, has been recently built to permit stereotactically guided open surgery, as well as the Gamma unit (Leksell 1971) for radiosurgery – which was in fact already possible with the cubic headframe.

The Todd-Wells System

The oval headframe of this system is attached to a semibase that allows X-ray precision, due to the fixed alignment. During surgery the target point is reached by moving the patient's head, fixed the headframe, guided by a x-y-z cursor, until the target point coincides with the center of the frame. The aiming device is a semiarc with a carrier for the instruments.

From this instrument are derived both the new BRW system (Brown 1979), already described, and the Kelly computer-guided system that will be the subject of further discussion in this book.

References

Bancaud J, Talairach J, Bonis A, Scaub C, Szikla G, Morel P, Bordas-Ferrer M (1965) La stéréo-électroencéphalographie dans l'epilepsie. Masson, Paris

Bancaud J, Talairach J, Geier S, Scarabin JM (1973) EEG et SEEG dans les tumeurs cérébrales et l'epilepsie. Edifor, Paris

Birg W, Mundinger F (1982) Direct target point determination for stereotactic brain operations from CT data and the calculation of setting parameters for polar-coordinate stereotactic devices. Appl Neurophysiol 45:387–395

Bosch DA (1986) Stereotactic techniques in clinical neurosurgery. Springer, Vienna
Brown RA (1979) A computerized tomography-computer graphics approach to stereotactic localization. J Neurosurg 50:715–720
Fox JL (1969) Selected readings in techniques of stereotactic neurosurgery. A bibliography through 1968. US Government Printing Office, Washington
Jacques S, Shelden CH, McCann GD, Freshwater DB, Rand R (1980) Computerized three-dimensional stereotactic removal of small central nervous system lesions in patients. J Neurosurg 53: 816–820
Kelly PJ, Alker J Jr, Goerss S (1982) Computer-assisted stereotactic laser microsurgery for the treatment of intracranial neoplasms. Neurosurg 10:324–331
Leksell L (1949) A stereotactic apparatus for intracerebral surgery. Acta Chir Scand 99:229–233
Leksell L (1971) Stereotaxis and radiosurgery. An operative system. Thomas, Springfield
Munari O (1987) Depth EEG in partial epilepsy. In: Engel J (ed) Surgical treatment of epilepsy. Raven, New York
Riechert T, Mundinger F (1955) Beschreibung und Anwendung eines Zielgerätes für stereotaktische Hirnoperationen. Acta Neurochir [Suppl] (Wien) 3:308–337
Scerrati M, Fiorentino A, Fiorentino M, Pola P (1984) Stereotaxic device for polar approaches in orthogonal systems. Technical note. J Neurosurg 61:1146–1174
Szikla G (1979) Stereotactic neuroradiology and functional neurosurgery: Localization of cortical structures by three dimensional angiography. In: Rasmussen T, Marino R (eds) Functional neurosurgery. Raven, New York, pp 197–217
Szikla G, Bouvier G, Hori T, Petrov V (1977) Angiotherapy of the human brain cortex: Atlas of vascular patterns and stereotactic cortical localization. Springer, Berlin Heidelberg New York Tokyo
Talairach J, Hecaen H, David M, Monnier M, De Ajuriaguerra J (1949) Recherches sur la coagulation thérapeutique des structures sous-corticales chez l'homme. Rev Neurol 81:4–24
Talairach J, David M, Tournoux P, Corredor H, Kvasina T (1957) Atlas d'anatomie stéréotaxique. Repérage radiologique indirect des noyaux gris centraux des régions mésencephalo-sous-optique et hypothalamique de l'homme. Masson, Paris
Talairach J, Szikla G, Tournoux P, Prossalentis A, Bordas-Ferrer M, Covello L, Jacob M, Mempel A, Buser P, Bancaud J (1967) Atlas d'anatomie stéréotaxique du télencéphale. Masson, Paris
Todd EM (1967) Todd-Wells manual of stereotactic procedures. Codman & Shurtleff, Randolph, Mass

A Computer Monitored Stereotactic CO_2 Laser System for Removal of CNS Intra-Axial Tumors: Technical Details and Clinical Results

P. J. KELLY

Introduction

Significant neurosurgical advances have been credited to new instrumentation such as the operating microscope, improved neuro-anesthesia, and the evolution of technical skill. Nevertheless, these advances have been of most benefit to patients with extra-axial lesions, meningiomas, acoustic neurinomas, aneurysms, and vascular lesions located on the surface or at the base of the brain. In fact, little progress has been made in the surgical resection of intracranial intra-axial neoplasms in over 50 years. Unless the lesion is situated in frontal, temporal, or occipital poles and treated by lobectomy, resections by internal decompression procedures of tumors located deep in the hemispheres, are usually insignificant cytokinetically. In addition, many surgeons consider tumors of the thalamus and basal ganglia inoperable.

There are good reasons for this failure to progress in the treatment of intra-axial lesions. First, surgeons have long depended on hand-eye coordination to guide surgical procedures. However, a surgeon's three-dimensional orientation decreases the deeper the procedure extends below the cortical surface. In addition, parts of primary intra-axial gliomas may not be significantly different in appearance from edematous brain tissue so that a surgeon may not be able to determine where the tumor ends and viable brain tissue begins. Hemispheric and deep-seated tumors are frequently located in neurologically important brain tissue. Therefore, it is not surprising that surgeons tend to be conservative with intra-axial neoplasms with the inevitable result that much of the lesion is left behind.

New imaging modalities such as CT scanning and more recently NMR scanning provide a three-dimensional data base which may represent the tumor as a three-dimensional volume in space. These data can be incorporated into a stereotactic coordinate system. The precision provided by the carbon dioxide laser can also be incorporated in a stereotactic surgical method for accurate removal of tumors located in space by these new imaging modalities (Kelly and Alker 1981; Kelly et al. 1982, 1983). The present report describes the technical details of such a system and our clinical experience using it.

Materials and Methods

Data Base Acquisition

Stereotactic CT Scanning

Patients undergo contrast enhanced CT scanning on a GE 8800 CT scanner with their heads fixed in a CT/MRI compatible stereotactic headholder utilizing a method described previously (Goerss et al. 1982). Attached to the headholder is a CT localization system which consists of 3 localizing devices on either side of the head and anteriorly. Each localization device contains 3 sets of one-fourth-inch carbon fiber bars (one oblique and 2 verticals). The localization system produces a series of 9 artifacts on each CT slice. The vertical height of the slice above the base ring is related to the distance of the middle (oblique bar) from the verticals on each localization device. These describe 3 points in space which define a plane.

Some patients also undergo stereotactic magnetic resonance imaging on a Picker .15 Tesla MRI unit. An MR localizing system similar to that used for CT scanning creates reference marks on each MR slice.

Following CT scanning and MR imaging, the patient is taken to the operating room and placed under general anesthesia. The archived data tapes from CT and MRI are read into a Data General Eclipse S140 computer system located in the operating room.

Tumor Volume Reconstruction

The CT and MRI slices are displayed on a Ramtek raster scan display terminal in the operating room. The central pixel of each of the 9 localization artifacts on each CT or MRI slice is automatically digitized by an intensity detection program. The surgeon than digitizes the outline of the tumor on each CT and MRI slice utilizing the terminal cursor and trackball subsystems. An interpolation program then creates intermediate slices at 1 mm intervals, and each slice is given a thickness of 1 mm. Thus, solid volumes in space are created for the CT and MRI defined limits of the tumor. In addition, these CT and MRI defined volumes may be sliced orthogonal to any specified surgical viewline.

The fiducial localization systems are directly related to the stereotactic frame positions. The computer calculates the mechanical adjustments necessary to bring a specified point within the center of the tumor volume into the focal point of the stereotactic frame, and the tumor volumes are reconstructed around that point and the surgical viewline is defined by arc and quadrant angle settings on the instrument.

Surgical Apparatus

In our arc-quadrant stereotactic system the patient's head moves with three degrees of freedom by means of a motorized X, Y, Z, slide to place an intracranial point in the center of a sphere described by arc and quadrant. A 400 mm double-gimballed

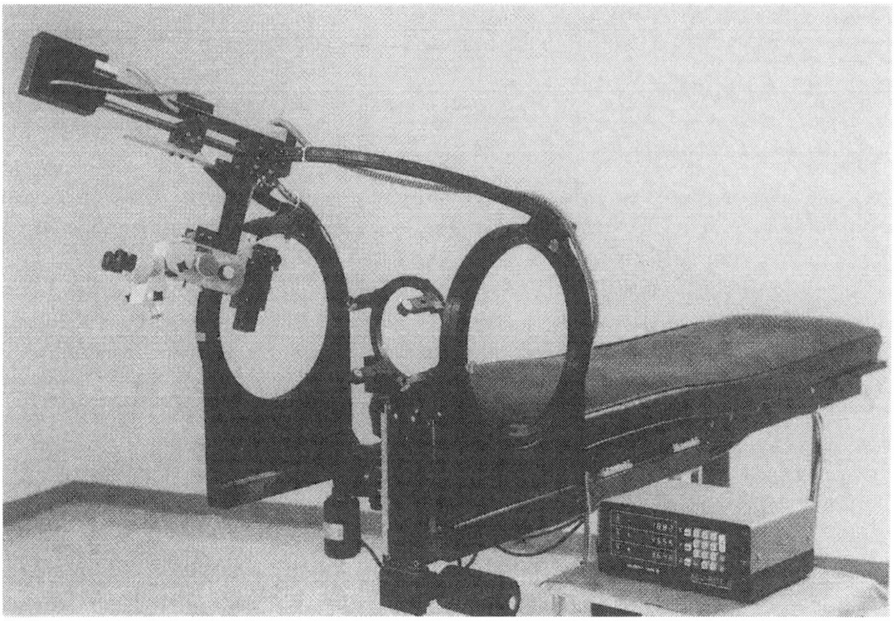

Fig. 1. Arc quadrant computer interactive stereotactic frame consisting of three-dimensional servomotor controlled slide system which moves patient's head in three-dimensions to place intracranial target in focal point of 400 mm arc which holds the operating microscope and laser manipulator system. Stereotactic coordinates are detected by optical encoders and displayed by a digital readout

arc is attached to the arc-quadrant stereotactic frame (Fig. 1). An operating microscope and laser micromanipulator box from a Sharplan 743 surgical laser are mounted on a moveable carriage which runs perpendicular to the arc's tangent. The laser and the microscope have 400-millimeter focussing and objective lenses respectively and are directed toward the focal point of the stereotactic frame. The distances of the microscope and laser carriage from the focal point of the frame are given by an optical digital read-out system.

The laser remote "joystick" is connected by a double-gimballed apparatus to 2 encoders which transmit the X, Y position of the laser beam to the computer. The position of the laser beam is displayed on a graphics terminal in the operating room as a cursor upon a calibration grid. Also displayed on the terminal are slices of the tumor cut orthogonal to the surgical approach and at the level at which the microscope and laser are focussed.

Surgical Procedure

The tumor is approached through nonessential brain tissue through a linear scalp incision, a 1½–2″ cranial trephine and circumferential dural opening. The white matter incision is made with the CO_2 laser and deepended as the laser and microscope are directed toward the focal point of the stereotactic frame. At the outer border of the tumor (the position of which is calculated along the viewline by the computer), the incision is undercut by reflecting the laser beam with a stainless steel mirror. A self-retaining stereotactic retractor is then inserted.

The tumor is vaporized slice by slice utilizing 65–85 watts of defocussed laser power in a continuous mode progressive from the most superficial slice to the deepest. The position of the laser beam is monitored by the surgeon on the computer graphics terminal in reference to the tumor outline which has been reformated from CT and MRI data. The reference stainless steel balls confirm the depth of the surgical procedure as they are encountered. In addition, serial AP and lateral teleradiographs are obtained during the procedure. Ultimately, a cavity is produced which conforms to sagittal and coronal reconstructions of the tumor from CT scanning data.

Clinical Results

70 patients have undergone 75 computer assisted stereotactic laser microsurgical procedures. The tumors were located in the thalamus in 8 patients, basal ganglia in 6 patients, corpus callosum in 4 patients, the deep frontal area in 2, frontoparietal area in 15 (right 10, left 5), deep in the parietal lobe in 10 (right 6, left 4), in the parieto-occipital area in 4 (left 2, right 2), in the occipital lobe in 4 (left 2, right 2), and in the temporoparietal area in 4 (left 3, right 1). 4 patients had lesions in the medial temporal area (2 on the left and 2 on the right). In addition, 4 patients had lesions in the third ventricle. One patient had a lesion resected from the midbrain, and 2 additional patients had lesions removed from the pons. 2 other patients had lesions within the lateral ventricle.

Histologically, there were 25 glioblastomas, 13 metastatic tumors, 14 grade 2 astrocytomas, 4 grade 3 astrocytomas, 2 oligodendrogliomas, and 6 arteriovenous malformations. 7 additional patients had miscellanous lesions including 2 ganglio-gliomas, 1 cavernous hemangioma, 1 fungal abscess, 2 meningiomas and 1 teratoma. 1 patient had an arteriovenous malformation within a grade 2 astrocytoma.

Neurologic condition was assessed 1 week postoperatively. 24 patients were found to have improved in their neurologic condition from their pre-operative level.

Patients were unchanged: 26 had been normal pre-operatively whereas 15 had a pre-operative deficit which did not improve postoperatively. 8 patients were worse following the procedure. 3 had developed a visual field defect following occipital approaches. (Of these 2 had dorsal posterior thalamic lesions and one, a posterior medial temporal lesion.) 1 patient remained in a lethargic vegetative state for several weeks following an approach to a midbrain lesion and then made an eventual though not complete recovery. 4 patients had worsening of a hemiparesis noted pre-operatively. There were 2 infections in this series: A fatal ventricular infection in a patient with a third ventricular teratoma who previously had been subjected to radiation and chemotherapy and had been deteriorating on high dose glucocorticoids pre-operatively and another patient who had also been treated with radiation and chemotherapy with a deep temporoparietal glioblastoma who developed a brain abscess postoperatively. The latter patient responded favorably to evacuation of the abscess and antibiotics.

Postoperative CT scans with contrast were obtained within 4 days following the surgical procedure. In 55 patients there was no apparent tumor on the postoperative

CT scan. 15 patients, however, had a small amount of residual tumor. 2 patients had re-operations within 3 months to remove residual tumor apparent on CT. One of these patients had had a large ganglioglioma which was filling the third ventricle. The other had a low-grade juvenile astrocytoma in the left basal ganglia. There were 3 procedures done for recurrent tumors. One patient had a re-operation for recurrent grade 3 astrocytoma in the right basal ganglia 1 year following the first operation, and another patient had 2 re-operations for recurrent meningiomas located in the left glomus of the choroid plexus. Re-operations were performed 1 year and 22 months following the first procedure. It was noted that re-operations could be performed easily utilizing the same incision, bone flap, and trajectory angles through the same white matter tract as during the first stereotactic procedure. It is believed that in certain cases, it is best to perform 2 operations on very large tumors rather than a single operation to avoid rapid decompression and intracranial vascular compromise.

14 of the 25 patients with glioblastomas have had recurrence of the tumor noted on follow-up CT scanning, usually between 3 to 12 months following this surgical procedure. The recurrent tumor is usually noted in the hypodense zone surrounding the contrast enhancing portion of the lesion which is resected, or as noted in 2 cases, generalized recurrence by seeding of the ventricular system was noted. 1 patient with a grade 3 astrocytoma of the posterior medial frontal area extending into the corpus callosum remains free of disease on CT scanning 4 years following surgery. So far, none of the 13 patients with metastatic tumors has had local intracranial recurrence following stereotactic laser resection of the tumor.

Discussion

The stereotactic technique has been used for years for placing a point into three-dimensional space for ablation of subcortical structures by stereotactic probes or for placement of deep brain stimulating electrode systems. Our technique carries stereotactic methodology further in that a three-dimensional volume is reconstructed in stereotactic space (Kelly et al. 1982; 1983). This volume may be reached with precision utilizing a stereotactically directed carbon dioxide laser system described here and in previous reports (Kelly and Alker 1981; Kelly et al. 1982; 1983). The laser beam is deflected by mirrors, and their mechanical linkage is controlled electronically. The position of the laser beam is input to the operating room computer for an on-line read-out of laser position in relation to planar tumor contours reformatted from stereotactic CT data.

Our postoperative neurologic results, utilizing this method, have been encouraging. The precision provided by the CO_2 laser coupled with our stereotactic technique results in more selective removal of the neoplasm with minimal trauma to surrounding brain tissue. The amount of damage done by the laser beam is related to the spot size of the laser with a thin zone of thermal necrosis on either side of the laser incision. Tumors in difficult areas, such as the thalamus can be removed without significant damage to surrounding brain tissue.

The surgical system is as accurate as the data base supplied to the operating room computer. We feel that our failure to significantly resect glioblastomas reflects the fact that the tumor extends beyond the limits suggested by CT scanning (Daumas-Duport et al. 1982). In the future, incorporation of other data bases, such as NMR, may increase the accuracy of our system for the resection of these lesions.

Summary

This report describes a system for the precision resection of deep seated intracranial neoplasms. A computer monitored stereotactically directed carbon dioxide laser is utilized to vaporize intracranial tumors volumes of which have been reconstructed from computed tomography and magnetic resonance data bases. The position of the laser is monitored on a computer graphics terminal in reference to planar slices through the tumor, reformatted orthogonally to the surgical approach. This procedure has been performed on 70 patients with satisfactory postoperative neurologic results. The system provides precise surgical control in 3 dimensional space for the safe precision resection of significant amounts of intra-axial neoplasms as demonstrated by postoperative contrast enhanced computed tomography.

References

Daumas-Duport C, Monsaingeon V, et al (1982) Serial biopsies: A double histological code of gliomas according to malignant and 3-D configuration, as an aid to therapeutic decision and assessment of results. Appl Neurophysiol 45:431–437

Goerss S, Kelly PJ, et al (1982) A computed tomographic stereotactic adaptation system. Neurosurg 10:375–379

Kelly PJ, Alker GJ Jr (1981) A stereotactic approach to deep-seated CNS neoplasms using the carbon dioxide laser. Surg Neurol 15:331–334

Kelly PJ, Alker GJ Jr, et al (1982) Computer assisted stereotactic laser microsurgery for the treatment of intracranial neoplasms. Neurosurg 10:324–331

Kelly PJ, Kall B, et al (1983) Precision resection of intra-axial CNS lesions by CT-based stereotactic craniotomy and computer monitored CO_2 laser. Acta Neurochir 68:1–9

An Advanced Therapeutic Approach to Actively Expanding Intracranial Cysts: Stereotactic Beta Endocavitary Irradiation of Craniopharyngiomas and Gliomas

C. Munari, A. Musolino, O. O. Betti, R. Clodic, S. Askienazy, and J. P. Chodkiewicz

Introduction

The existence of an actively expanding cyst represents one of the most common complicating factors of many intraaxial lesions: in all these lesions the complete surgical removal is theoretically the most suitable therapeutic approach. Unfortunately, several topographic localizations as well as many anatomo-clinical reasons make the realization of this optimal solution difficult. More precisely, two different kinds of intracranial pathology offer evident examples of such difficulties: the cystic component of craniopharyngiomas (as it is largely referred to) (Miles 1977; Mori et al. 1980) and the (pseudo) cysts of the low grade gliomas (Szikla and Peragut 1975).

Even taking into account the obvious and well known differences between these two types of lesions, some of their anatomo-clinical aspects allow us to consider the "cyst problem" as a unique problem, from the therapeutical point of view. In fact: a) the craniopharyngiomas are rarely exclusively solid (10%–30% of different studies) (Banna 1976; Bond et al. 1965; Braun et al. 1982; Hoff and Patterson 1972; Petito et al. 1976) and, in many cases, the bad clinical evolution is strictly linked to the expansion of the cystic component; b) in low grade gliomas, the cystic component may be responsible for the clinical symptomatology, above all when the solid part is small and the cystic part is actively expanding.

The recurrency of both, craniopharyngioma and glioma cysts, often makes a definitive surgical recovery impossible.

For these reasons, since 1953 many authors have attempted to treat the craniopharyngioma cysts by endocavitary irradiation (Leksell and Liden 1953; Wycis et al. 1954).

Analogous attempts in glioma cysts have been carried out since 1973 by Szikla and Peragut at St. Anne Hospital (Szikla and Peragut 1975). The aim of this paper is to show the different procedures we used stereotactically treating the cysts satellite of these two types of lesion by endocavitary irradiation.

Patients

A. Craniopharyngiomas

We treated 16 patients (8 males, 8 females) aged from 3 to 63 years (mean 21 y.). The first clinical symptomatology appeared 1 to 21 years (mean 6 y.) before their admission to the St. Anne hospital.

11 of them had been operated on before. All of them had an important impairment of visual acuity and 15 of them had a limitation of the visual field. Five of the 6 children (10 years) had important endocrine disturbances. Diabetes was demonstrated in 7/16.

The neuropsychological examination showed important memory disturbances in 4 and an impairment of the IQ in another 4. The neurological examination showed a slight unilateral motor deficit in 4, and a facial paralysis in 2.

B. Low Grade Gliomas Cysts

20 patients (12 males, 8 females, aged from 5 to 53 years, mean 26 + 12) having 24 cysts were treated.

The first clinical symptomatology of the tumor appeared at 4 months to 18 years (mean 5 years) before the admission to our Department.

The clinical signs, directly linked to the expanding cysts, appeared 4 to 19 months (mean 9 + 4) before their admission. 14/20 presented partial epileptic seizures, representing the most important part of the clinical symptomatology in 6 cases. 5 presented a lateralized motor deficit; 2 sensitive disturbances; 2 visual deficit, 2 diskynesias. One patient had clinical signs of intracranial hypertension. The anatomical localization was in the dominant hemisphere in 11 patients (central 3; parietal 2; fronto-temporal 1; basal ganglia 5), and in the non-dominant in 8 (frontal 1; central 3; temporal 1; occipital 1; basal ganglia 3).

4/20 had been operated on before, 11 had received irradiation (3 external radiotherapy; 8 interstitial curietherapy).

Methodology

A. Diagnostic Study

1. Stereotactic Neuroradiological Approach

The spatial definition of the cerebral structures was realized according to the methodology of Talairach (Talairach et al. 1975; Talairach et al. 1967). In practical terms, after the fixation of Talairach's stereotactic frame to the skull of patients, we obtain in stereoscopic conditions (Szikla et al. 1977) the arteriography and the ventriculography. The data furnished by the CT scan was "manually" reconstructed on the stereotactic intracranial space.

Serial stereotactic biopsies permitted us to:

- define exactly the type of lesions;
- appreciate (when necessary) the spatial configuration of the lesion(s);
- obtain a direct visualization of the cysts (cystography with air and positive contrast).

2. *Volume Definition*

A precise definition of the cyst's volume is the unavoidable basis of any intracavitary treatment. Three different methods were used:

- the necessarily rough estimation furnished by the CT scan images was made in all patients;
- the more precise volumetric data obtained by direct visualization (cystography) was evaluated in all patients;
- in some patients with irregular shaped cysts we also calculated the cystic volume by isotopic dilution method (Backlund 1973; Netzeband et al. 1983).

a) Craniopharyngiomas
In 3 patients, the lesion was essentially cystic; in 6, the solid part constituted 1/3 of the global lesion; in 7, the two components were practically similar. The cyst's volume varied from 1 cc–270 cc (mean 51 cc).

b) Low Grade Gliomas
In 4 patients, we were unable to appreciate a solid volume: the gliomas were essentially cystic. In 9, the solid/cystic ratio was 1/3; and in 7, the volume of the two components was apparently similar. The cyst volume varied from 20 cc–240 cc (mean 85).

B. Therapeutic Approach

1. *Choice of the Isotopes*

Since the first paper by Leksell and Liden (1953), several authors reported good results of intracystic injections of colloidal P 32, Au 198 and Y 90 preparations in cystic craniopharyngiomas.

The comparative evaluation of the physical properties of the 3 nontoxic colloidal isotopes (Au 198, Y 90, Re 186) (Table 1), as well as the good results obtained in our

Table 1

Radio-isotopes	186 Re	198 Au	90 Y
Physical half-life (h)	90.6	64.8	64.8
Maximum beta energy (MeV)	1.076	0.966	2.284
Mean beta energy (MeV)	0.358	0.322	0.761
Maximum range in soft tissue (mm)	4.5	4	10.9
Mean value depth in soft tissue (mm)	1	0.87	2.9
Major gamma energy (MeV)	0.137	0.412	–

Department since 1973 (Musolino et al. 1985; Schaub et al. 1979; Szikla and Peragut 1975; Szikla et al. 1984) are the reasons for the use of the Re 186 in 13/16 patients with cystic craniopharyngiomas and in 19/20 patients with low grade cystic gliomas. In the remaining patients we used the Au 198 three times and the Y 90 once.

2. Endocavitary Treatment

According to Loevinger et al. (1956), the quantity of mCi which must be injected in the cyst is defined on the basis of several data: the type of the colloidal isotope, the volume of the cyst, the length of the irradiation, the wall dose to be delivered.

After the stereotactic localization of the cyst with the Talairach system, we first verify the imperviousness of the cyst-wall with only tracer doses (300 μCi–500 μCi; 0.5 cm–1 cm) of Re 186. The endocavitary treatment is realized 1–2 weeks after the test proving the absence of leakage. The colloidal solution (0.5 cm–5 cm) is stereotactically injected in the cyst, after the aspiration of 1 cc–6 cc of the cystic fluid.

In patients with a craniopharyngioma, the Re 186 injected varied from 5–100 μCi (mean 28 μCi). The cyst-wall varied from 16,000 to 83,000 rads (mean: 43,000 rads).

In patients with gliomas we injected 9 μCi–80 μCi (mean 35.5 + 21), delivering a cyst-wall-dose varying from 11,200 to 191,000 rads (mean 42,000).

Intraoperative and then daily scintigraphic controls were performed during the treatment. At the end of the treatment (5–14 days), the cysts were evacuated.

In cases of major leakage to CSF spaces, the cyst was immediately evacuated.

3. Controls

A cystography (air and positive contrast) is realized during the same intervention permitting the evacuation of the cyst. A CT scan control is always realized before the discharge of the patients, and then, every 6 months during the first 18 months, and then, every year. In one patient who died 54 months after the endocavitary treatment, the autopsy data was obtained.

Results

A. Anatomoclinical Results

1. Craniopharyngiomas

One patient, suffering from a severe endocrine deficiency, died 2 weeks after the treatment. This patient had been operated on by a classical approach 2 years before, and was then submitted to an external radiotherapy (50 Gys). Another patient, coming from a foreign country, did not come back for clinical controls.

Thus, we can only report the results concerning 16 cysts treated in 14 patients (one patient had 3 cysts). These patients were periodically controlled for 20–80 months (mean follow-up 44 months).

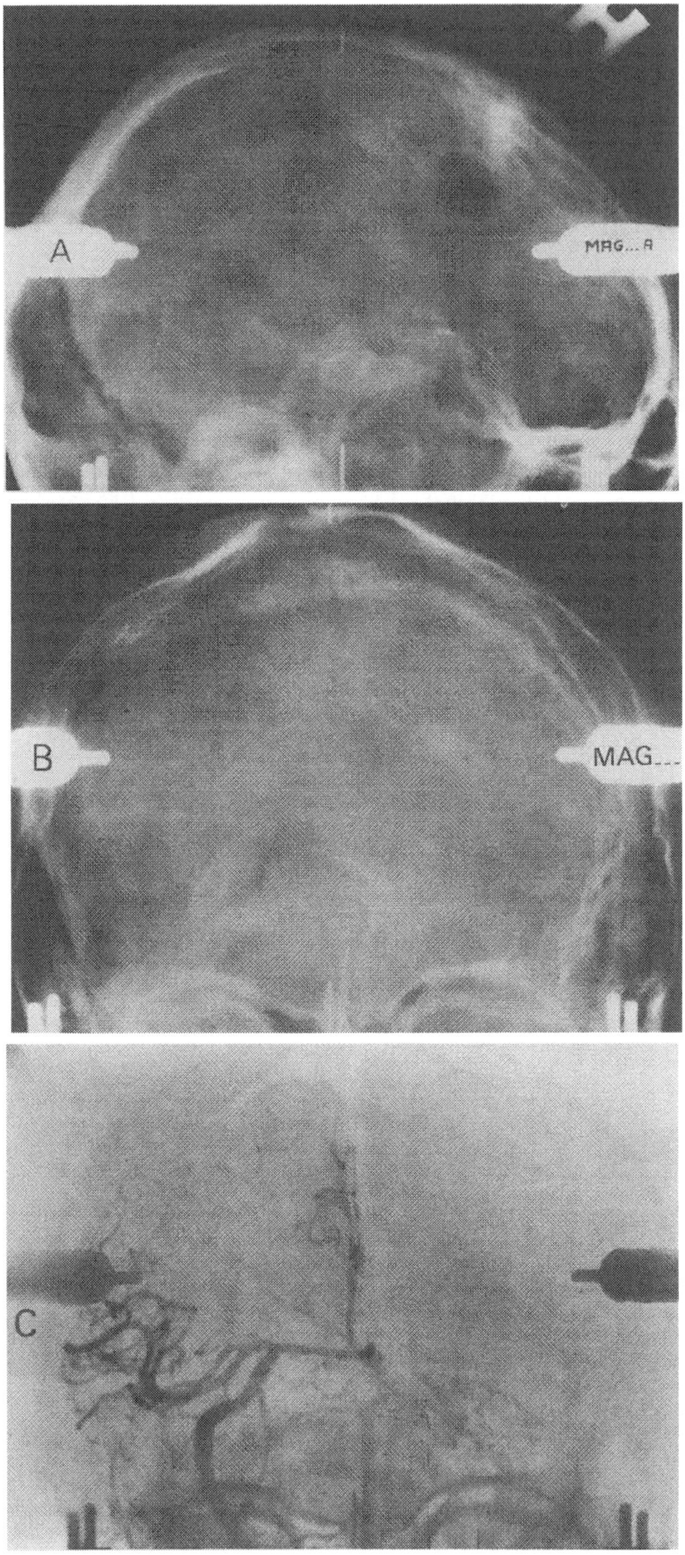

Fig. 1A–C

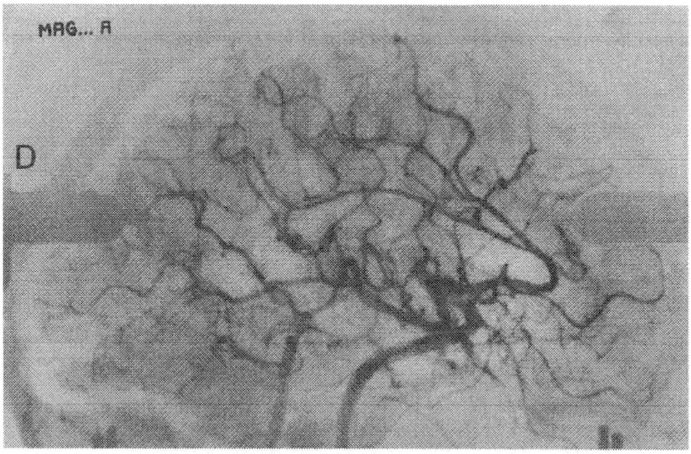

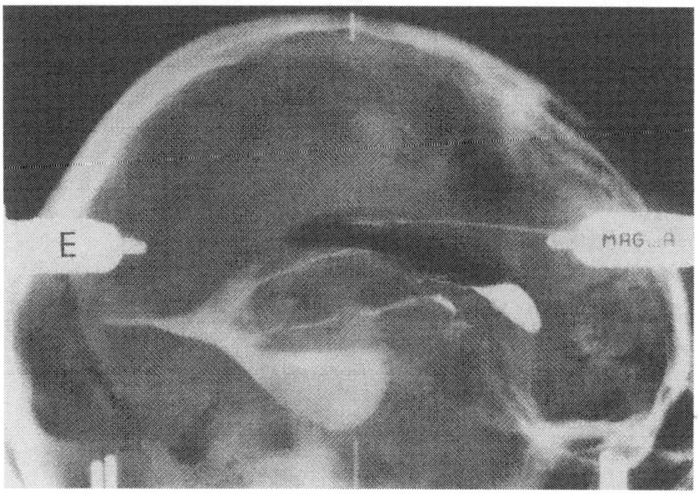

Fig. 1D, E

Fig. 1. A Evolution of a cyst of craniopharyngioma after the beta endocavitary irradiation: **A, B** Stereotactic skull X-rays showing a very large cyst wall calcification. **C, D** Carotid angiography in stereotactic conditions: note the displacements of the carotid, the anterior and the middle cerebral arteries. **E** Stereotactic ventriculography by frontal approach. The picture is obtained of a sitting patient. Note the modifications of the frontal horn, of the 3rd ventricle, of the right temporal horn, and of the aqueduct. **F** CT scan before the treatment. **G** CT scan 49 months after the treatment. (Part-figures **F** and **G** appear on p. 126)

12 cysts completely disappeared 2–24 months after the treatment (Fig. 1A–G). The other 4 had a volume reduction varying from 65%–30% of the initial volume. 2 cysts required a double therapeutic injection, 7 and 11 months after the first endocavitary treatment. Among the patients with an important loss of visual acuity, only 1 presented a dramatic improvement (from 1/10 to 10/10) without changes in the visual field (follow-up: 6.2 years).

3 of the 4 patients with a slight diminution of visual acuity, had a very good improvement (5.5/10 to 9.5/10). The fourth with an optic atrophy was impaired.

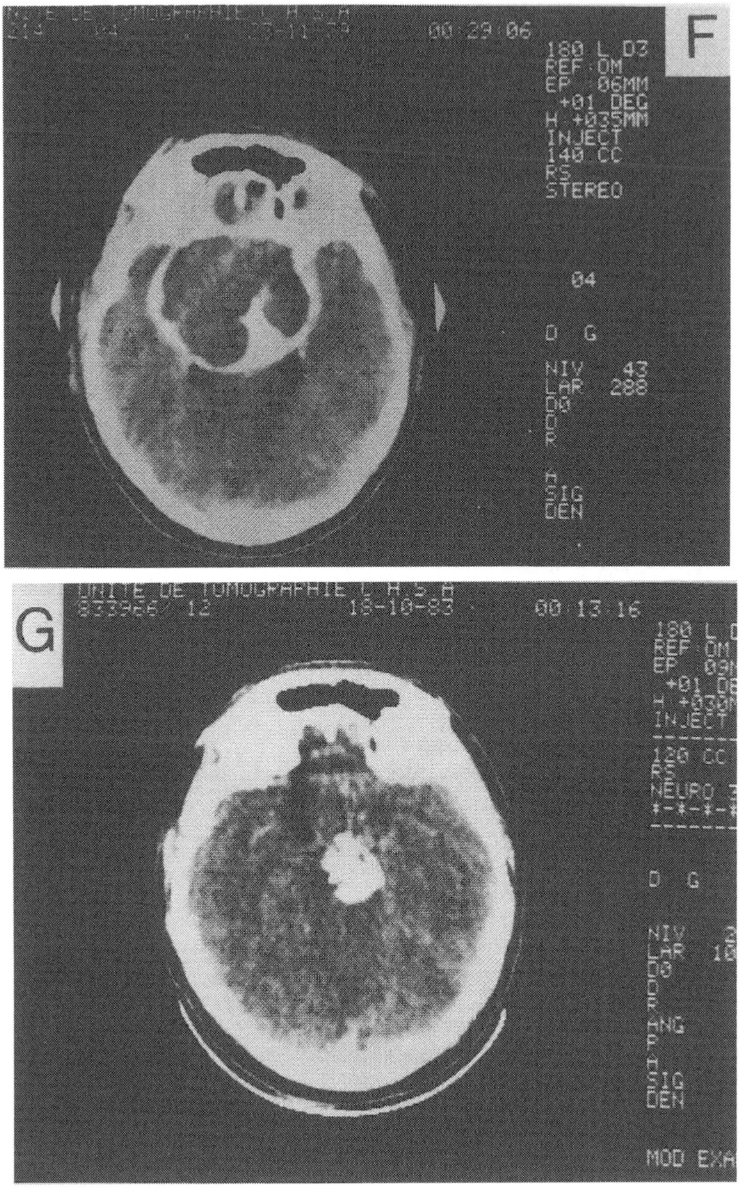

Fig. 1F, G

The memory disturbances, pre-existing in 4, completely and progressively disappeared (1 year) in all of them. We never observed any impairment of the memory in the others. A slight improvement was observed in 3 of the 9 patients with an endocrine insufficiency.

2. Gliomas

11 patients had a grade I glioma (pilocytic 9; cerebellar 1; ganglioglioma 1) and 9 had a grade II glioma (fibrillar 4; protoplasmic 4; oligo-astrocytoma 1).

The follow-up of the 24 cysts varied from 1–10 years (m: 35 months).

6 patients (5 with grade II and 1 with grade I glioma) died 12–54 months after the endocavitary treatment. In the patient who died 59 months after the treatment, the CT scan showed the complete disappearance of the cyst at 30 months: the autopsy confirmed the absence of the cyst. In 3 the cyst was inactivated, in 1 the cyst volume was 60% of the initial one.

9 cysts were inactivated and stabilized: the volume did not rise after the treatment, but was not less than 80% of the initial cyst. The mean follow-up of these cysts is 18 months.

2 cysts had a reduction of 50% and of 70% 3 years after the treatment.

13 cysts completely disappeared (Fig. 2A, B, C). Their mean follow-up is 37 months. The disappearance was first shown by the CT scan after a period of 2–30 months (mean: 18 months).

B. Side Effects

1. Leakages

a. Craniopharyngiomas
Out of a total of 44 injections (24 as test; 20 therapeutics) we recorded 8 leakages (4 tests; 4 therapeutics) representing 18%. No clinical effects were observed.

b. Gliomas
Out of a total of 48 injections (24 tests; 24 therapeutics) we observed 1 leakage (2%), 48 hours after a therapeutic injection. The cyst was immediately evacuated, without clinical effects. 3 months later the test and the treatment were repeated without problems.

2. Late Recurrence

A late re-expansion of the cyst was observed in 1/16 cyst of craniopharyngiomas (6%) 11 months after the treatment. This phenomenon was also observed in 3/24 cysts of low grade gliomas (12%), 12, 24 and 50 months after the treatment.

A new treatment permitted us to obtain the disappearance (in two) or the retraction (in the other two) of these 4 cysts.

3. Edema

This type of side effect was never observed in patients with craniopharyngiomas.

Considering the 27 therapeutic injections (24 + 3 repeated) in patients with gliomas, the occurrence of a white matter edema, in 4, represents less than 15%. In

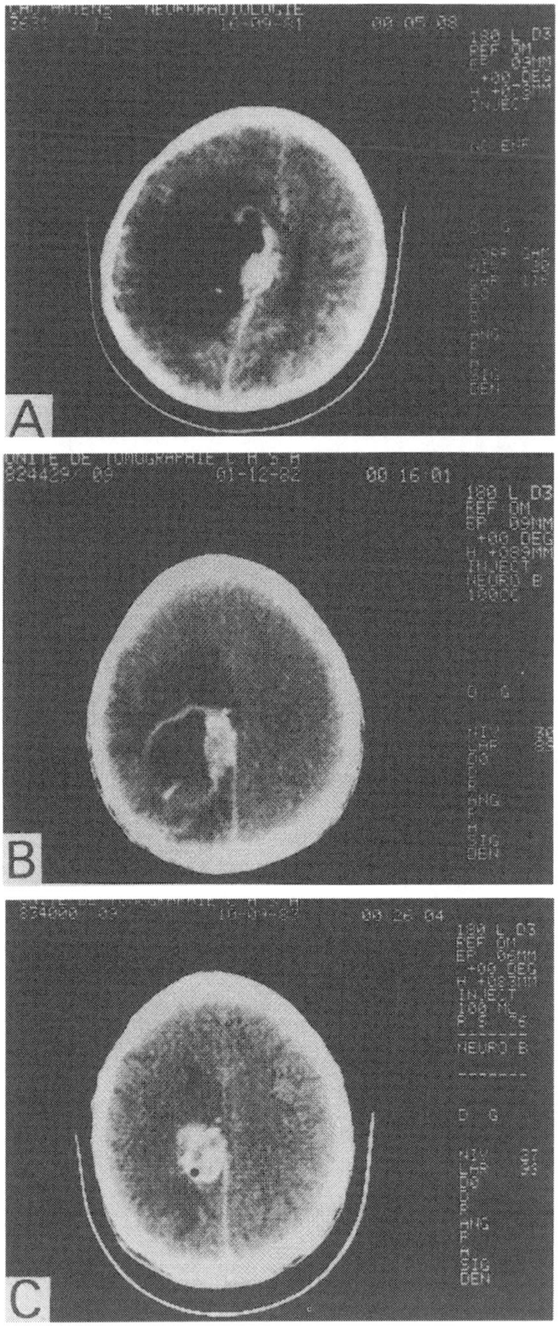

Fig. 2A–C. Evolution of a glioma cyst after the beta endocavitary irradiation. **A** Before the treatment. **B** 9 months after the treatment. **C** 18 months after the treatment fact, this complication, with transitory aggravation of neurological symptoms, occurring 1–2 months after the treatment, concerns only 4 out of the 13 injections giving a cyst wall dose more than 400 Gy (30%). The absence of edema after the 18 injections giving a cyst-wall dose less than 400 Gy is statistically significant ($P = 0.001$).

Discussion

The problem of the recurrence of the cysts of craniopharyngiomas may be considered as a strictly surgical problem (Garcia-Uria 1978; Gordy et al. 1949; Matson and Crigler 1969; Pertuiset 1975; Petito et al. 1976), radically improved by the utilization of the microscope (Hoff and Patterson 1972; Mori et al. 1980).

In fact, the experience of many surgical teams shows that the risk of functional damage is relatively high (Cavazzuti et al. 1983; Mori et al. 1980), and that the recurrence of the cyst remains possible even after an apparently complete removal (Shapiro et al. 1979). The practical usefulness of direct evacuations of the cysts is doubtful because of the high number of early or late recurrences (Cushing 1926; Hoffman et al. 1977; Russel and Pennybaker 1961).

Thus, considering the risk of open surgery, the refilling of the cyst after the aspiration, the not generally admitted effectiveness of external radiotherapy, we have used since 1975 the endocavitary treatment of the craniopharyngioma cysts (Szikla and Peragut 1975).

This therapeutic approach is based on the stereotactic methodology of Talairach and it is only slightly modified in comparison with Szikla's methodology (Szikla et al. 1984).

The choice of the Re 186 is related to the following facts:

- the Beta emission has a mean penetration of 1 mm, thus avoiding secondary lesions in neighboring structures;
- soft gamma emission (0.137 MeV) permitting easy scintigraphic controls;
- relatively short physical half life (90.6 h);
- limited secondary necrosis (1 mm–1.5 mm);
- rare, only dose-dependent, white matter edema.

The second choice isotope may be the Au 198, especially in those craniopharyngiomas with a large cyst wall: the Au 198 has a beta mean penetration similar to the Re 186, associated with a relatively high gamma emission (0.412 MeV), three times more than the Re 186, inducing more extended necrosis and edema. We prefer not to use the Y 90 (Julow et al. 1985), because of the high beta-penetration, double that of the Re 186, and of the absence of gamma-emission, not permitting an adequate control of the possible leakages. This undesiderable effect occurred in 18% of the injections in craniopharyngioma cysts and only in 2% of the injections in glioma cysts. This important difference is probably due to the different anatomical organization of the cysts in these two types of pathology: the glioma cyst is in fact a "pseudo"-cyst, with a completely different organization of the wall. As far as the treatment of the glioma cysts was concerned, we stopped treating those satellite of high grade gliomas, the treatment benefit being rapidly compensated by the solid tumor progression. We now limit the endocavitary treatment only to the really expanding cysts of low grade gliomas (especially grade I). In these indications, the results seem to be very encouraging, with 62% of important retraction of the cysts (complete disappearance in 54%) and a long term stabilization in the remaining ones.

In the first patient of this series, the cyst wall doses were very variable: our actual experience shows that cyst-wall doses of 500 Gy may permit us to obtain very good results without side effects.

Finally, we want to stress that this type of treatment needs a complex technological equipment including:

- the stereotactic Talairach system (permitting stereoscopic studies),
- the possibility of safe use and stock radio-isotopes,
- informatic instruments to correctly calculate the dosimetry,
- gamma scintigraphic cameras to control the possible leakages,
- possibility of isolating the treated patients.

Acknowledgements. We want to express our gratitude to Odile Missir (M.D.) for the precision of her C.T. controls and to Catherine Daumas-Duport (M.D.) for her diagnostic per-operative help.

References

Backlund EO (1973) Studies on craniopharyngiomas. III. Stereotactic treatment with intracystic Yttrium 90. Acta Chir Scand 139:237–247

Banna M (1976) Craniopharyngioma: based on 160 cases. Br J Radiol 45:206–223

Bond WH, Richards D, Turner E (1965) Experiences with radioactive gold in the treatment of craniopharyngioma. J Neurol Neurosurg Psychiatry 28:30–38

Braun JF, Pinto RS, Epstein F (1982) Dense cystic craniopharyngiomas. Am J Neuroradiol 3:139–141

Cavazzuti V, Ficher EG, Welch K, Belli JA, Winston KR (1983) Neurological and psychophysiological sequelae following different treatments of craniopharyngioma in children. J Neurosurg 59:409–417

Cushing H (1926) Studies in intracranial physiology and surgery. Oxford University Press, London

Garcia Uria G (1978) Surgical experience with craniopharyngioma in adults. Surg Neurol 9:11–14

Gordy PD, Peet MM, Kahn EA (1949) The surgery of craniopharyngiomas. J Neurosurg 6:503–517

Hoff JT, Patterson RH (1972) Craniopharyngiomas in children and adults. J Neurosurg 26:299–302

Hoffman HJ, Hendrick B, Humphreys RP, Buncic JR, Amstrong GL, Jenkin RDT (1977) Management of craniopharyngioma in children. J Neurosurg 47:218–227

Julow J, Lanyi F, Hajda M, Simkovics M, Arany I, Toth S, Pasztor E (1985) Acta Neurochir 74:94–99

Leksell L, Liden K (1953) A therapeutic trial with radioactive isotopes in cystic brain tumor. In: Radioisotope techniques, vol 1: Medical and physiological applications. Her Majesty's Stationery Office, London, p 76

Loevinger R, Tapha EM, Brownell GL (1956) In: Hine GJ, Brownell GL (eds) Radiation dosimetry. Academic Press, New York

Matson DD, Crigler JF (1969) Management of craniopharyngiomas in childhood. J Neurosurg 30:377–390

McMurry FG, Hardy RW Jr, Dohn DF, et al (1977) Long term results in the management of craniopharyngiomas. Neurosurgery 1:238–241

Miles J (1977) Pump drainage: a palliative manoeuvre for the treatment of craniopharyngiomas. J Neurol, Neurosurg Psychiatry 40:120–126

Mori K, Handa H, Murata T, Takeuki J, Mina S, Osaka K (1980) Results of treatment for craniopharyngioma. Child Brain 6:303–312

Musolino A, Munari C, Blond S, Betti O, Lajat Y, Schaub C, Askienazy S, Chodkiewicz JP (1985) Traitement stereotaxique des kystes expansifs de craniopharyngiomes par irradiation endocavitaire beta (Re 186, Au 198, Y 90). Neurochirurgie 31:169–178

Netzeband G, Sturm U, Georgi P, Sinn H, Schnabel K, Schlegel W, Schabbert S, Marin-Grez MS, Gahbauer H (1983) Results of stereotactic intracavitary irradiation of cystic craniopharyngiomas. Comparison of the effect of Yttrium 90 and Rhenium 86. 6th Meeting of the European Society for Stereotactic and Functional Neurosurgery, Rome, June 1983

Pertuiset B (1975) Craniopharyngiomas. In: Vinken PJ, Bruyn GW (eds) Handbook of clinical neurology, vol 18. North-Holland, Amsterdam, pp 531–572

Petito CK, De Girolami U, Earle KM (1976) Craniopharyngiomas. A clinical and pathological review. Cancer 37:1944–1952

Rougerie J, Fardau M (1962) Les craniopharyngiomes. Masson et Cie, Paris, p 217

Russel RW, Pennybaker JB (1961) Craniopharyngioma in the ederly. J Neurol Neurosurg Psychiatry 24:1–13

Schaub C, Bluet-Pajot MT, Videau-Lornet C, Askienazy S, Szikla G (1979) Endocavitary beta irradiation of glioma cysts with colloidal 186 Rhenium. In: Szikla G (ed) Stereotactic cerebral irradiations. Elsevier North-Holland Biomedical Press, Amsterdam New York Oxford, pp 293–302

Shapiro K, Till K, Grant DN (1979) Craniopharyngiomas in childhood: a rational approach to treatment. J Neurosurg 50:617–623

Szikla G, Peragut JC (1975) Irradiation interstitielle des gliomes. In: Constans JP, Schliener M (eds) Radiotherapie des tumeurs du systeme nerveux central. Neurochirurgie [Suppl 2] 21:187–228

Szikla G, Bouvier G, Hori T, Petrov V (1977) Angiography of the human brain cortex. Atlas of vascular patterns and stereotactic cortical localization. Springer Verlag, Berlin Heidelberg New York, pp 1–237

Szikla G, Musolino A, Miyahara S, Schaub C, Askienazy S (1984) Colloidal rhenium 186 in endocavitary beta irradiation of cystic craniopharyngiomas and active glioma cysts. Long term results, side effects and clinical dosimetry. Acta Neurochir suppl 33:331–339

Talairach J, Szikla G, Tournoux P, Prossalentis A, Covello L, Bordas-Ferrer M, Jacob M, Mempel E (1967) Atlas d'anatomie stéréotaxique du téléncéphale. Masson ed., Paris, p 323

Talairach J, Peragut JC, Fanarier P, Manrique M (1975) The role of stereotaxic exploration in neurosurgical interventions. In: Salamon G (ed) Advances in cerebral angiography. Springer Verlag, Berlin

Wycis HT, Robbins R, Spiegel-Adolf M, Meszaros J, Spiegel EA (1954) Treatment of a cystic craniopharyngioma by injection of radioactive P32. Conf Neurol 14:193–202

Stereotactic Brachytherapy of Gliomas

CHR. B. OSTERTAG

Introduction

Although many surgeons claim that by using microscopic techniques they can reach virtually all compartments in the brain, others consider this approach mechanical, particularly in cases of intrinsic tumors in functionally critical areas. Radiotherapy is therefore preferred as it is the least traumatic treatment for "nonresectable" tumors.

Current radiotherapeutic practice seeks to optimize the therapy by careful definition of the target volume, the dose distribution, and the quality of radiation. The use of stereotactic techniques in conjunction with modern imaging and radiation techniques has aroused new enthusiasm for the stereotactic approach to brain tumors. Considerable progress has been made in two areas: more precise beam techniques and the placement of radioactive sources into tumors, i.e. brachytherapy.

Radiation Principles and the Biology of Gliomas

A brain tumor is usually radiated with a maximum dose. The dose limit is set by the surrounding tissue and by the critical structures, such as the hypothalamus, in the neighboring area. In conventional radiation therapy, the tolerance of the healthy brain is critical. This implies a therapeutic ratio, i.e., a difference in radiosensitivity between healthy brain and brain tumor tissue.

Generally speaking, the cells most sensitive to radiation exposure, are the ones that are still capable of replication. In the healthy brain, these include glial elements, particularly in younger age groups, and endothelial cells in all age groups. The susceptibility of the individual cell is dependent on its position within the cell cycle at the time of exposure. Radiosensitivity is usually greatest in the mitotic phase and early DNA synthesis phase with less sensitivity in the resting phase (Zeman 1968; Caveness 1980).

It is estimated that 0.1% of endothelial cells are in the mitotic phase at any given time (Fowler and Dennekamp 1977). Consequently, in the healthy brain the first radiation changes on the capillary endothelium are found when the tolerance of the brain is exceeded (Rubinstein 1972).

Growth characteristics and mitotic activity of gliomas are major determinants for the biological effect of therapeutic radiation. We know from clinical experience that glioblastomas can grow very rapidly and that, on the other hand low grade astrocytomas, like cerebellar astrocytomas or optic chiasm gliomas, grow very slowly, if at all. The studies on the cell kinetics of various gliomas by Hoshino et al. (1972) and Hoshino (1984) have shown that the labelling index as an indicator for the mitotic activity is indeed very high in medulloblastomas (12%) and glioblastomas (9.3%), is moderate in anaplastic astrocytomas (4%), and is very low in fibrillary astrocytomas (0.8%). The mitotic activity in pilocytic astrocytomas is in the range of that of the healthy capillary endothelium.

Astrocytomas have a very limited proliferating capacity. Therefore, in well differentiated gliomas it is important to remove as much tumor as possible (Hoshino 1984; Laws et al. 1984). Many patients with low grade gliomas, especially those with cerebellar astrocytomas, survive for several years after incomplete surgical resection because even partial removal gives the residual tumor space to grow and reduces the number of cells in the proliferating pool.

With these biological data in mind it seems likely that conventional irradiation will be effective in proliferating tumors, i.e., malignant tumors, such as glioblastomas and anaplastic astrocytomas. Conventional radiotherapy will not be effective in tumors with a very low proliferation capacity, in which cases it seems wiser to remove as much tumor as possible. When, however, there is no harmless way of surgical removal, as in cases of low grade astrocytomas in the basal ganglia, brainstem, or in functionally critical cortical areas, radiosurgery in the form of low dose rate brachytherapy is considered the appropriate treatment modality. Biologically, the local radionecrosis achieves a kinetically significant reduction of the tumor cellular burden.

Principles of Brachytherapy

Leksell (1971) pointed out the difference between radiotherapy and radiosurgery. By focussing the radiation dose into very small volumes, either by gamma-units or by the implantation of radioactive material, one can effect a predetermined volume of radionecrosis. At present, iodine-125 (I-125) and iridium-192 (Ir-192) are the most suitable radioisotopes with which to produce a tumor radionecrosis. In order to avoid conflict with radioprotection regulations, both sources of low activity are used as lost implants or removable high activity sources. Iodine-125 permanent implants are presently preferred. The half life of 60.4 days makes it possible to economically stock sources with different activities.

Radiation from an interstitially implanted permanent source such as iodine-125 is continuously delivered at very low dose rates (20 cGy/h) compared with dose rates delivered by external beam. The therapeutic ratio is also enhanced as a result of the rapid dose fall-off in tissue within a distance of millimeters. Since radiation from iodine-125 sources is effectively attenuated by interjacent tissue such as brain, bone of the skull, and scalp, no special radioprotection is necessary for either patient or medical personnel.

Experimental Brachytherapy

How does the normal brain tissue, which is usually involved in interstitial tumor irradiation, tolerate this radiation? To study this question, the sequential morphological changes with permanently implanted iodine-125 and iridium-192 sources in healthy dog brains were observed up to one year after implantation (Ostertag et al. 1982; 1983; 1984a; Janzer et al. 1985).

Due to the low photon energy of iodine-125, much of the energy is absorbed by the tissue next to the implant. Accordingly, around the implanted seed we found a necrosis which was always calcified and which was observed as early as 25 days after the implantation. The size of the necrosis did not increase further after 70 days–90 days. The dose which accumulated later apparently did not contribute to the necrotizing effect. The total size of the necrosis depends on the activity implanted.

Iridium-192 differs from iodine-125: it has a gamma radiation energy which is ten times higher than that of iodine-125. The iridium-192 implants effected delineated liquifying and calcifying necrosis. The transitional zone was characterized by extensive gliosis, which was demonstrated by GFAP positive astrocytes and numerous blood vessels with an enlarged diameter and thickening of the vessel wall. Blood vessels with radiation damage could not be detected in the animals surviving more than 70 days, which means, that blood vessels either were damaged and incorporated into the necrosis or that they were not affected. Demyelination was always restricted to the ipsilateral white matter and showed no progression at later stages of intersitial radiation.

In no case did we observe the extensive leuko-dystrophic myelin break down which occasionally develops as a late complication of external radiotherapy. Vasogenic edema seems to be the most important single factor for local demyelination.

Vasogenic edema, i.e., the permeability changes attributed to the radionecrosis, was also investigated as a part of this complex radiolesion having profound clinical implications. For this, we used quantitative autoradiography with a C-14 alpha-amino-isobutyric acid (Groothuis et al. 1985). This test substance has many properties which make it ideal for measuring rates of blood-to-tissue transport. It is a synthetic, inert, neutral amino acid of low molecular weight that slowly crosses the blood brain barrier. It is taken up rapidly by brain cells but is not metabolized, and therefore can serve as an ideal tracer of unidirectional transport for radiation-induced capillary permeability disturbance (Blasberg et al. 1983).

In all of the dogs, regardless of time of exposure, which ranged from 7 days–717 days, we found a clearly demarcated sphere of blood brain barrier break down around the implant, which can persist for over one year following insertion of the I-125 seed.

In a further experimental series avian-sarcoma-virus-induced dog brain tumors were used as a model for interstitial irradiation of neoplastic tissue (Ostertag et al. 1984b). These avian-sarcoma-virus induced dog brain tumors can serve as a primary model for human brain tumors, since these tumors are autochthonous. Their blood supply and growth are characteristic of a primary brain tumor. The features of these tumors closely resemble those of human anaplastic astrocytomas. The intratumoral placement of iodine-125 seeds in anaplastic gliomas produced sharply defined calcifying necroses with unaffected vital tumor tissue outside the necrosis. The calcified necrosis was complete after 90 days when the transitional zone was no longer detectable. When compared with the necrosis in healthy brain tissue, the volume of necrosis in neoplastic tissue was approximately 3–5 times greater. That means that there is a differential necrotizing effect of iodine-125 in normal and neoplastic brain tissue.

Clinical Brachytherapy

Clinical Dosimetry

With the experimental data in mind, the dosimetry for permanent I-125 implants is calculated on the basis of tumor shape as indicated by CT and MR and verified bioptically by stereotactic serial biopsies. In clinical practice, the implantation of radioactive material is usually carried out immediately following the stereotactic serial biopsy, provided that the morphological diagnosis is clear and unequivocal. Angiography of cerebral vessels, preferably under stereotactic conditions, is a prerequisite for each procedure (Ostertag et al. 1980).

The clinical dosimetry for permanent iodine-125-implants is carried out with respect of the physical and biological properties: iodine-125 emits primarily a low energy X-ray radiation of 27 keV–35 keV. According to several investigators, the "relative biological efficiency" (RBE) seems to be in the range of 1.2–1.4 (Sondhaus 1981). The specific dose rate factor used for dosimetry is 1.32 cGy per h and mCi at 1 cm in tissue. For the clinical situation, the calculated accumulated dose for permanent implants ranges between 3500 cGy and 7000 cGy. This dose is calculated to accumulate within 90 days on the outer radius of a tumor.

Clinical Results and Discussion

In our clinical practice, brachytherapy is routinely used as a primary treatment for low grade gliomas after biopsy. Occasionally, it is also used to irradiate residual tumors after incomplete surgical removal, and it is used as a combination of external beam radiation and local radiotherapy for high grade gliomas.

Since the opening of the department (April 1984), 34 patients have been treated with permanent iodine-125 implants. All of the patients exhibited progressive neurological deficits or had evidence of CT-controlled tumor growth or both. The majority were astrocytic tumors (23 cases) or mixed oligo-astrocytomas (8 cases), which were located in the thalamus-basal ganglia or in the temporal and frontal lobes (Tables 1, 2).

The calculated accumulated doses ranged between 2000 cGy for a pontine astrocytoma and 7000 cGy for a temporal lobe tumor (mean 5500 cGy). The radius of irradiated tumors varied from 8 mm to 20 mm (mean 15.5 mm).

The implants produced favorable responses resulting in stabilization (14 patients) or improvement of neurological symptoms (13 patients), which was paralleled by a CT controlled reduction in tumor mass (Fig. 1). 4 patients with anaplastic astrocytomas (WHO III) and 2 patients with extended oligoastrocytomas (WHO II) re-

Table 1. Brachytherapy: diagnoses of tumors

Tumor type (WHO)	No. of cases
Astrocytoma I	6
Astrocytoma II	12
Oligo-astrocytoma II	8
Anaplastic astrocytoma III	5
PNET[a]	2
Pituitary adenoma[b]	1
Total	34

[a] Recurrences after previous external beam radiotherapy
[b] Third recurrence after resection and external beam radiotherapy

Table 2. Brachytherapy: localization of tumors

Tumor site	No. of cases
Basal ganglia-thalamus	11
Hypothalamus	2
Temporal lobe – insula	12
Frontal lobe	5
Midbrain – pons	3
Pituitary	1
Total	34

ceived additional external beam radiotherapy with doses up to 4000 cGy. In 3 patients with anaplastic astrocytomas, the combined irradiation did not halt clinical deterioration. These patients died within 3, 4, and 6 months, respectively. 2 patients received brachytherapy for recurrent primitive neuro-ectodermal tumors of the pineal region after previous external beam radiotherapy. Both have responded well so far (observation times 2 months and 1 year, respectively). There have been no complications or death related to the operative procedure of implanting seeds. The clinical data of this new series are considered premature and data on survival are currently meaningless.

The larger Freiburg series, which now comprises more than 500 cases, quotes 5 year survival rates of 54% for pilocytic astrocytomas, 45% for astrocytomas grade II (WHO), and 4% for anaplastic astrocytomas. The survival rate for oligodendrogliomas is 34% after 5 years. These figures represent, of course, very inhomogenous groups (Mundinger et al. 1979; Ostertag et al. 1981).

Combined interstitial and external radiotherapy was mainly developed by the Paris group with the late Gabor Szikla (Szikla et al. 1981; 1984). They used both

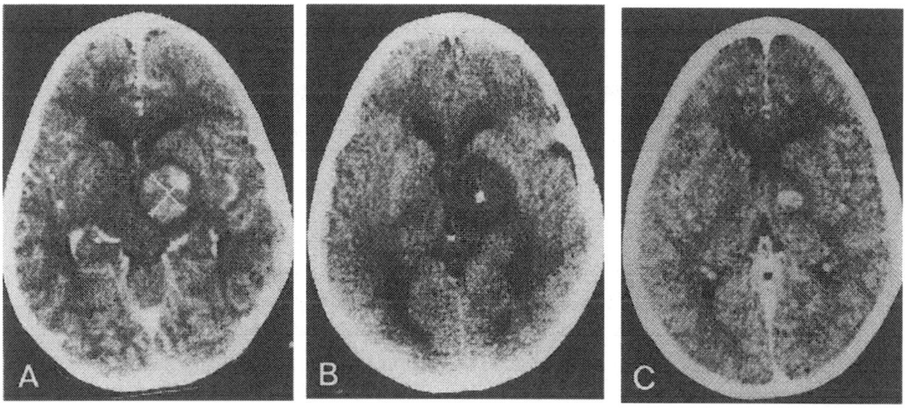

Fig. 1A–C. Stereotactic brachytherapy for a basal ganglia astrocytoma. **A** CT demonstrates a delineated pilocytic astrocytoma in the basal ganglia of a 10 year-old boy, who exhibited progressive spasticity of the left extremities. **B** 2 iodine-125 seeds with a total activity of 4.2 mCi were implanted, which delivered a calculated accumulated dose of 6000 cGy on the outer diameter of the tumor. The seeds are located in the center of the tumor. **C** A CT-control one year later revealed a considerable shrinkage of the tumor with mineralization around the implant. The motor deficit has disappeared and the boy is going well at school

removable high activity sources for local irradiation of the tumor volume and external radiation for the greater target volume which covers a wider area. That allows the application of more moderate doses compared with external beam radiotherapy alone. Survival data published by the Paris group (Szikla et al. 1984) show a 78% survival for gliomas grade I, 69% for gliomas grade II, 55% for gliomas grade III, and 19% of gliomas grade IV. It is understood, however, that these groupings follow a different classification scheme and represent different treatment modalities. Rougier et al. (1984) quote a 50% survival rate after 5 years for anaplastic gliomas treated with combined temporary interstitial and external beam radiotherapy. Gutin et al. (1984), and Kelly et al. (1978) also report favorable survival of patients with malignant gliomas. To date, however, no survival rates are available.

Brachytherapy with low dose rate permanent implants is not without hazards. Complications result mainly from overdosage, i.e., neurological deterioration can result from secondary effects of vasogenic edema or direct radiation damage of functionally critical structures (Kiessling et al. 1984). Unwanted side effects can be avoided with careful dosimetry. Nausea, vomiting, and somnolence syndrome, however, commonly seen after external cranial irradiation, are not observed after interstitial radiotherapy.

To compare interstitial radiotherapy with conventional fractionated beam therapy is a difficult undertaking. Prospective controlled trials on brain tumors have yet to be carried out. Clinical results of both external beam radiotherapy and interstitial radiotherapy have been based on studies of populations of patients grouped by tumor name only, regardless of the origin of the tumor, the age of the patient, the clinical presentation and course, and without a commonly accepted classification scheme. The only way to obtain more conclusive figures is through a multicenter, multimodality treatment study. Various centers in Europe are currently working on setting up such a study.

Conclusions

Advanced technology can only be used advantageously when the underlying biological phenomena are understood. The thin margin between tumor sensitivity and healthy brain sensitivity to radiation is a considerable limitation of conventional radiation therapy. Furthermore, because of the heterogeneity of gliomas on the one hand, and the proliferating characteristics of most gliomas on the other hand, radiotherapy and chemotherapy are relatively ineffective, when indicated on the basis of general recommendation.

On the basis of biological, experimental, and clinical data an individualized treatment for each patient is desirable. External radiation therapy is recommended for highly malignant, rapidly growing tumors which are infiltrating or are known to metastasize. Tumor specific or even individual tumor cell-type specific chemotherapy may be more effective in the future. Brachytherapy with I-125 permanent implants is recommended for slowly proliferating, differentiated, non-resectable tumors in functionally critical areas. It enables the surgeon to achieve a radiosurgical tumor removal while carefully avoiding radiation damage to the normal surrounding brain. It has been the authors personal experience for many years now that brachytherapy with iodine-125 permanent implants is effective for local tumor control, and is, at the same time, the least traumatic treatment when carefully applied.

References

Blasberg RG, Fenstermacher JD, Patlak CS (1983) Transport of alpha-aminoisobutyric acid across brain capillary and cellular membranes. Journal of Cerebral Blood Flow and Metabolism 3: 8–32

Caveness WF (1980) Experimental observations: Delayed necrosis in normal monkey brain. In: Gilbert HA and Kagan AR (eds) Radiation damage to the nervous system. Raven Press, New York

Fowler JF, Denekamp J (1977) Radiation effects on normal tissues. In: Becker FE (ed) Cancer – A comprehensive treatise. Plenum Press, New York, pp 139–180

Groothuis DR, Wright DC, Ostertag Ch B (1985) Sequential changes of capillary permeability in normal canine brain associated with I-125 interstitial radiotherapy. J Neurosurg (in press)

Gutin PH, Philips TL, Wara WM, Leibel SA, Hosobuchi Y, Levin VA, Weaver KA, Lamb S (1984a) Brachytherapy of recurrent malignant brain tumors with removable high-activity Iodine-125 sources. J Neurosurg 60: 61–68

Gutin PH, Bernstein M (1984b) Stereotactic interstitial brachytherapy for malignant brain tumors. Prog Tumor Res, vol 28. Karger, Basel, pp 166–182

Hoshino T (1984) A commentary on the biology and growth kinetics of low-grade and high-grade gliomas. J Neurosurg 61: 895–900

Hoshino T, Barker M, Wilson Ch B, Boldrey EB, Fewer D (1972) Cell kinetics of human gliomas. J Neurosurg 37: 15–26

Janzer RC, Kleihues P, Ostertag Ch B (1986) Early and late effects on the normal dog brain of permanent interstitial Iridium-192 irradiation. Acta Neuropathol (Berl) 70: 91–102

Kelly PJ, Olson MH, Wright AE (1978) Stereotactic implantation of Iridium-192 into CNS neoplasms. Surg Neurol 10:349–354

Kiessling M, Kleihues P, Gassega E, Mundinger F, Ostertag ChB, Weigel K (1984) Morphology of intracranial tumors and adjacent brain structures following interstitial Iodine-125 radiotherapy. Acta Neurochirurgica [Suppl] 33:281–289

Laws ER, Taylor WF, Clifton MB, Okazaki H (1984) Neurosurgical management of low-grade astrocytoma of the cerebral hemispheres. J Neurosurg 61:665–673

Leksell L (1971) Stereotaxis and radiosurgery. An operative system. Charles C Thomas, Springfield, Ill

Mundinger F, Busam B, Birg W, Schildge J (1979) Results of interstitial Iridium-192 brachy-curie therapy and Iridium-192 protracted long term irradiation. In: Stereotactic cerebral irradiation, INSERM Symposium No 12. Elsevier/North Holland Biomedical Press, Amsterdam, pp 303–319

Ostertag Ch B, Mennel HD, Kiessling M (1980) Stereotactic biopsy of brain tumors. Surg Neurol 14:275–283

Ostertag Ch B, Mundinger F, Weigel K (1981) Biopsie stéréotactique et radiotherapie interstizielle des tumeurs cérébrales. Med et Hyg 39:1994–2008

Ostertag Ch B, Hossmann KA, Kerckhoff W v d (1982) Radiation effects of Iridium-192 implants in the cat brain. Nucl-Med, Band XXI/Heft 3

Ostertag Ch B, Weigel K, Warnke P, Lombeck G, Kleihues P (1983) Sequential morphological changes in the dog brain after interstitial Iodine-125 irradiation. Neurosurgery 13/5:523–528

Ostertag Ch B, Groothuis D, Kleihues P (1984a) Experimental data on early and late morphologic effects of permanently implanted gamma and beta sources (Iridium-192, Iodine-125 and Yttrium-90) in the brain. Acta Neurochirurgica [Suppl] 33:271–280

Ostertag Ch B, Warnke P, Kleihues P, Bigner D (1984b) Iodine-125 interstitial irradiation of virally induced dog brain tumors. Neurol Res 6:176–180

Rougier A, Pigneux J, Cohadon F (1984) Combined interstitial and external irradiation of gliomas. Acta Neurochirurgica [Suppl] 33:345–353

Rubinstein LJ (1972) Radiation changes in intracranial neoplasms and the adjacent brain. In: Tumors of the central nervous system, 2nd edn. Armed Forces Inst of Pathology, Washington, DC

Sondhaus CA (1981) I-125: Physical properties, photon dosimetry and effectiveness. In: George FW (ed) Modern interstitial and intracavitary radiation cancer management. Masson, New York

Szikla G, Betti O, Szenthe L, Schlienger M (1981) L'experience actuelle des irradiations stéréotaxiques dans le traitment des gliomes hemispheriques. Neurochirurgie 27:295–298

Szikla G, Schlienger M, Blond S, Daumas-Duport C, Missir O, Miyahara S, Musolino A, Schaub C (1984) Interstitial and combined interstitial and external irradiation of supratentorial gliomas. Results in 61 cases treated 1973–1981. Acta Neurochirurgica [Suppl] 33:355–362

Zeman W (1968) The effects of atomic radiation. In: Minckler J (ed) Pathology of the nervous system, vol 2. McGraw-Hill, New York, pp 864–939

Progress in Interstitial Irradiation of Inoperable Malignant Brain Tumors

H. W. Pannek, F. Oppel, M. Brock, and H. Ernst

In recent years marked advances have become evident in the treatment of inoperable semimalignant gliomas (I–II WHO) which can be regarded as curative.

In this context interstitial irradiation – and specifically the permanent implantation of 192 Iridium and 125 Iodine – has proven to be very effective. Survival times of 12 years or more, with a 5-year survival time of 50%, have been reported (Mundinger 1983). Complete remissions of 2 years have so far been observed by us (Oppel et al. 1985).

On the other hand, permanent or temporary implantation alone does not provide successful results in cases of malignant gliomas (III–IV) and metastases. Here the combination with percutaneous irradiation is recommended (Gutin et al. 1984; Hochberg and Pruitt 1980). Percutaneous irradiation with or without chemotherapy constitutes an alternative form of treatment in this type of tumor. However, this form of therapy only lengthens the mean survival times to 28 weeks when total brain irradiation of 50 Gy is applied and to 42 weeks at 60 Gy as compared to 18 weeks in untreated patients (Walker et al. 1979).

An attempt has been made to improve the disappointing results by increasing the total dose to as much as 80 Gy in some cases (Salazar et al. 1979). Despite the high tumor dose, local recurrences persisted without any appreciable increase of survival time.

Attempts with radiosensitizer likewise showed no additional therapeutic effect (Sack et al. 1982). Having in mind the above-mentioned drawbacks, and with the aim of increasing the therapeutic efficiency, we have devised an improvement of the afterloading technique (Mundinger 1969) which consists in (1st) *fractionated* afterloading and (2nd) its association with percutaneous irradiation. This new approach was developed in close cooperation with and under the auspices of the department of radiation therapy of the Klinikum Steglitz.

This modified procedure offers the following advantages:

1. The spatial dose distribution can be adjusted to the size and shape of the tumor.
2. A relatively high radiation dose can be delivered to the center of the tumor (spatially superselective).
3. Fractioning and/or discontinuation of radiotherapy can be elected and selected freely.
4. Undesirable protracted effects of irradiation are safely avoided.
5. High radiation volume is ensured thanks to the association with percutaneous irradiation.

Whatever the type of therapy applied, its purpose must be to increase survival time and to improve life quality in view of the fact that it can only be palliative.

Since 1980, 21 patients with inoperable deep seated brain tumors were subjected to stereotactic biopsy and interstitial irradiation at the Department of Neurosurgery of the Universitätsklinikum Steglitz taking into consideration the above mentioned viewpoints. Of these tumors 15 were situated in the midbrain and 6 in the brain stem. Permanent 192 Iridium implantation was performed in 18 cases, while fractionated afterloading as devised by us was applied in 3 patients.

Methods

Calculation of tumor shape, volume, target points and dosimetry were based on the angiograms, computerized tomograms and MR-images – when available. In the cases of permanent 192 Iridium implantation, marginal tumor dose was limited to 6000–8000 rad in the brain stem area and to 8000–10000 rad in the midbrain area. Biopsy and isotope implantation were performed by conventional stereotactic technique and with the aid of an adapted biopsy forceps, a disposable metallic implantation tube specifically developed by us for the technique of fractionated afterloading was employed. The distal deep tip of this tube is blind while the proximal (superficial) end is open so as to receive the radiation source (as if it were a needle guide). Precision of positioning of both the tube and the seeds by this technique is of the order of ±1 mm.

To establish diagnosis and check target determination prior to implantation, 4–6 biopsies are made along the tube track and histologically classified at once. The center of the tumor is the target in cases of permanent implantation of 192 Iridium while for the afterloading procedure the tip of the metal tubing is placed at the center of the distal tumor periphery. Thus the metal tubing is inserted along the longitudinal axis of the tumor. The metal tube has an outer diameter of 3 mm. Since it must remain implanted for up to 10 days, it is firmly fixed in a threaded bolt screwed into the burrhole (Fig. 1). The following steps are necessary:

1. Drilling of the burrhole with a stereotactically guided drill.
2. Cutting an M-10-winding into the walls of the burrhole.
3. Inserting the threaded bolt previously adjusted to the thickness of the cranium.
4. Stereotactically introducing the metal tube through the threaded bolt until the target is reached.
5. Fixing the metal tube with a setscrew.
6. Radiologic control of the position of the metal tube.
7. Cutting the metal tube immediately above the threaded bolt using a cutting device.
8. Closing the skin.

The threaded bolt is built so as to permit satisfactory skin closure (Fig. 2).

The patient can move freely with the implanted metal tube. The radiation source is driven into the metal tube by an automatic arm of the radiation unit. Irradiation is

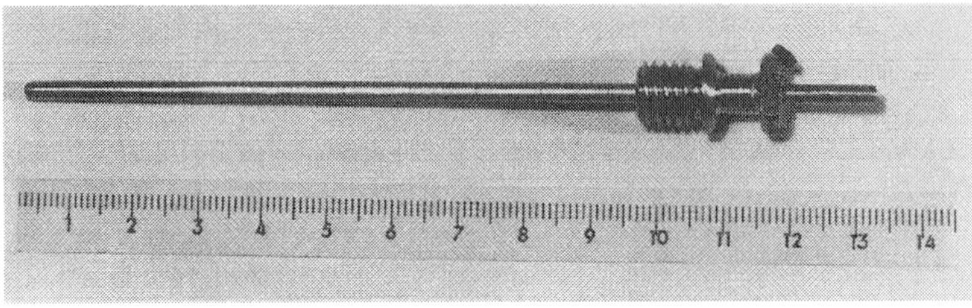

Fig. 1. The metal tubing and the thread-bolt (measurement in cm's)

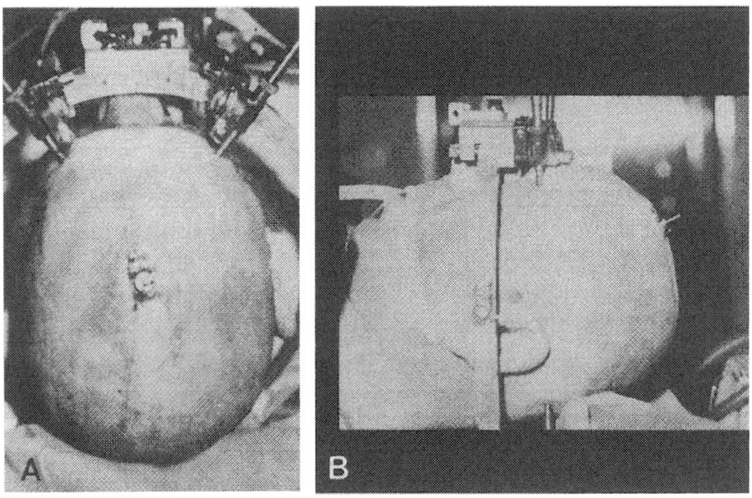

Fig. 2. A Patient fixed in the stereotactic ring. Skin closure is performed after insertion and fixation of the catheter in the screw bolt. **B** Lateral view showing the protruding catheter for later introduction of the radiation source

performed once daily for 3 to 4 minutes as calculated by computer according to the dose to be applied. Only during the short irradiation procedure must the patient remain in a special room as required by the regulations for radiation protection (Fig. 3). The total radiation dose is fractionated into up to 6–8 individual doses, the marginal tumor dose being limited to approximately 3000 rad. After conclusion of the radiation cycle the metal tube and bolt are easily removed under local anesthesia. Following an interval of 3 weeks percutaneous irradiation "boost" of 3000 rad is initiated subsequent to renewed CT studies. The order in which percutaneous and interstitial irradiation should be administered has not been established yet. Further experience is required to understand the interaction of these two forms of therapy.

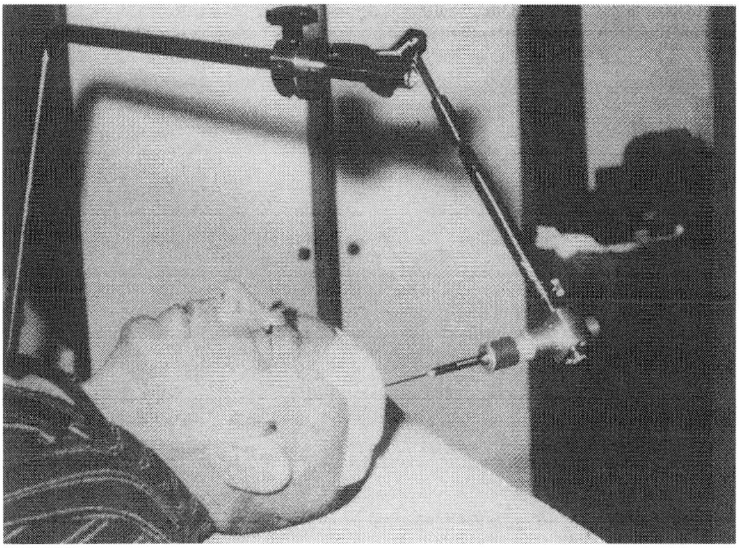

Fig. 3. Patient in a special room as required by the regulations for radiation protection

Results

All patients submitted to permanent implantation of 192 Iridium experienced a regression of neurological deficits to varying degrees. In 4 cases computerized tomography revealed a complete remission during a follow-up period of 1 to 2 years (Table 1). In 4 cases tumor growth could not be checked. Age, sex, tumor size, marginal tumor dose and survival rates are shown in Table 1. Results must be considered preliminary and are probably still influenced by histology.

Of the 3 patients treated by fractionated afterloading the first had an anaplastic astrocytoma (stage III) of the left basal ganglia reaching the callosum (Fig. 4) while the second and third patients had astrocytomas (stage III) in the sensory-motor cortex. 6 single doses were applied in cases 1 and 2 and 8 single doses in case 3. 24 h were chosen as the therapy interval. The patients did not evidence any neurological alterations attributable to radiation.

Discussion

At present interstitial irradiation through permanent 192 Iridium implantation must be regarded as the most effective therapy for inoperable deep seated brain tumors. As compared to percutaneous irradiation, the high radiation dose in the center of the tumor and the steep decrease towards the surrounding tissue constitute a clear advantage. However, permanent implantation of radionuclides has the disadvantage that temporal dose distribution cannot be optimized. In addition, protracted irradia-

Table 1. Histological diagnosis, maximal tumor spread, radiation activity, survival time of the dead *(N)* and the still living patients *(S)*, cases of fractionated afterloading – first results

Pat. no.	Age (years)	Sex	Pathology	Diameter max. (mm)	Activity (mCi)	Survivors (S), non-survivors (N)	R/RR
Brain-stem							
1. J.Z.	32	M	Astrocytoma I	55	4.2	S	R
2. G.S.	70	M	Metastatic Ca.	30	1.3	S	
3. F.B.	59	M	Astrocytoma II	60	5.5	N	
4. H.G.	40	M	Astrocytoma III	40	2.1	N	R
5. N.H.	5	M	Oligodendroglioma II	60	3.3	S	
6. R.Y.	37	F	Astrocytoma II	40	2.1	S	RR
Basal-ganglia							
1. H.D.	57	M	Oligodendro-astrocytroma II–III	50	4.6	S	R
2. D.Z.	46	M	Oligodendroglioma II	40	4.0	N	
3. G.H.	33	M	Astrocytoma II	40	1.6	S	RR
4. G.L.	26	M	Astrocytoma I	40	2.0	S	RR
5. B.B.	45	F	Astrocytoma II–III	40	1.6	N	R
Thalamus							
1. H.Sch.	54	F	Astrocytoma III–IV	50	3.4	N	
2. G.L.	67	F	Astrocytoma I	30	1.5	S	
3. A.K.	30	F	Astrocytoma III	30	1.6	N	
4. H.G.	22	M	Astrocytoma I	40	3.0	S	RR
5. B.H.	26	F	Oligodendro-astrocytoma III	50	3.2	S	R
6. H.D.	35	F	Astrocytoma II	40	3.1	S	RR

R = recidive, RR = remission.

tion must be accepted and constitutes an undesirable side effect both for biological reasons and due radiation protection. As a general statement it must be said that interstitial irradiation is of value only in cases of low-malignancy gliomas (I–II), while external irradiation is indicated for high-malignancy gliomas (III–IV).

Although percutaneous irradiation (external beam) also permits fractionation, a superselective high dose irradiation restricted to the tumor cannot be attained by this method, not even with the multiple-field or pendulum technique. Nevertheless, we believe that the advantages of both methods can be combined. Varying the position of the radiation source within the metal tube permits a precise tailoring of the isodoses to the shape and size of the tumor (Fig. 5). The patient can move freely and the stereotactic procedure does not involve any radiation exposure for the personnel. The technique of implantation as developed by us is reliable and reproducible. To optimize positioning of the metal tube a mathematical program was developed which provides a 3-dimensional tumor reconstruction directly from the stored CT-informa-

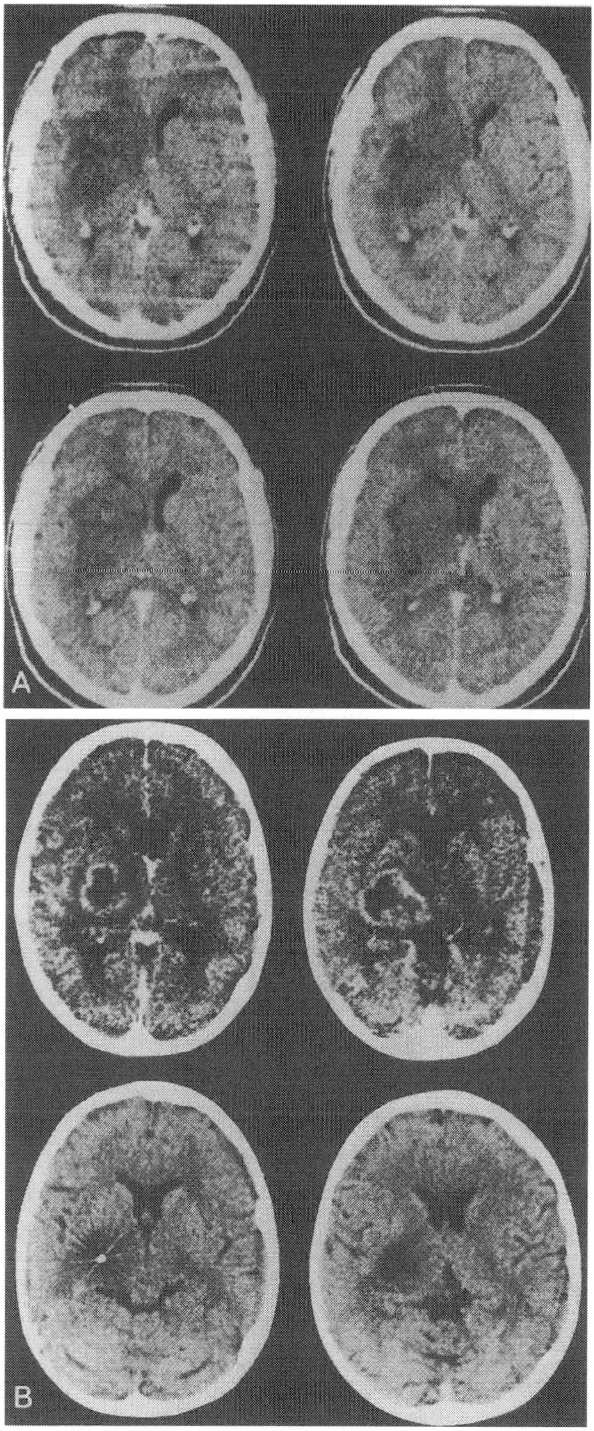

Fig. 4A, B. CT of the astrocytoma III of the left basal ganglia. **A** Before the afterloading irradiation. **B** After 2 months of the afterloading irradiation

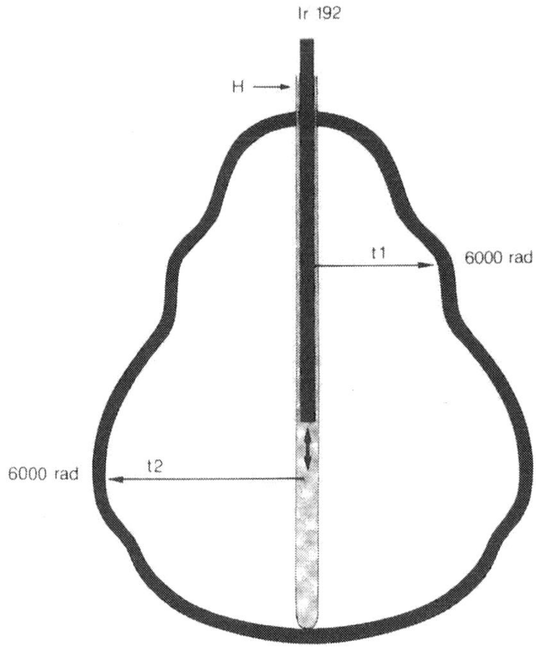

Fig. 5. Schematic representation of a tumor with irregular contours. A catheter *(H)* has been introduced through the longitudinal axis of the tumor for interstitial irradiation by fractionated afterloading. The radiation source *(Ir)* remains *(arrow)* at each point of the tumor for the individual length of time *(t 1–2)* required, the purpose being to attain a marginal tumor dose of *6000 rad* through different durations of exposure

tion and allows on-line plotting of the isodose curves. Even though no long-term results are available as yet, the conception of volume-steered fractionated tumor irradiation appears to constitute an important step forward in the treatment of malignant brain tumors. Local post-radiation brain edema can be treated with dexamethasone.

Both the permanent implantation of isotopes and the transient insertion of the metal tube have not led to aggravation of clinical symptoms in the patients treated so far. As concerns late results of our technique, the available data are still insufficient to permit evaluation. The main scope of this paper was the presentation of a new technique to improve efficiency and safety of irradiation and to decrease discomfort for the patient.

Note: Since conclusions of this manuscript one of the patients submitted to fractionated afterloading by the technique here described developed an abcess. This patient died despite abcess evacuation by puncture. Autopsy was refused.

References

Gutin WH, Phillips TL, Wara WM, Leibel SA, Hosobuchi Y, Levin VA, Weaver KA, Lamb S (1984) Brachytherapy of recurrent malignant brain tumors with removable high-activity iodine-125 sources. J Neurosurg 60:61–68

Hochberg FH, Pruitt A (1980) Assumptions in the radiotherapy of glioblastoma. Neurology 30: 907–911

Mundinger F (1969) Erfahrungen mit der stereotaktischen interstitiellen Brachytherapie mit IR-192 Gamma-Med bei infiltrierenden Hirntumoren. Fortschr Röntgenstr 110:254–261

Mundinger F (1983) Stereotaktische intrakranielle Bestrahlung von Tumoren mit Radioisotopen (Curie-Therapie). In: Dietz H, Umbach W, Wüllenweber R (eds) Klinische Neurochirurgie. Thieme, Stuttgart New York, pp 519–565
Oppel F, Pannek HW, Brock M (1985) Die interstitielle Bestrahlung maligner inoperabler Hirntumoren nach dem Prinzip des "after loading". Psycho 11:422–424
Sack HA, Calcanis H, Godehardt E, Weidtmann V, Zülch KJ, Ammon J, Bamberg M, Herbst M, Keim H, Kleibel F, Makoski HB, Potthoff PC, Schlegel G, Schnepper E (1982) Die postoperative Strahlenbehandlung von Astrozytomen Grad 3 und 4 mit dem Strahlensensibilisator Misonidazol. Strahlentherapie 158:466–469
Salazar OM, Rubin P, Feldstein ML, Pizzutiello R (1979) High dose radiation therapy in the treatment of malignant gliomas: final report. Int J Radiat Oncol Biol Phys Vol 5, 10:1733–1740
Walker MD, Strike TA, Sheline GE (1979) An analysis of dose-effect relationship in the radiotherapy of malignant gliomas. Int J Radiat Oncol Biol Phys Vol 5, 10:1725–1731

Radiosurgery of Arterio-Venous Malformations

O. O. Betti, V. E. Derechinsky, and D. H. Galmarini

Introduction

Radiosurgery is the method based on Leksell's principle of cross-firing X-ray beams.

After Steiner's (Steiner et al. 1972) pioneering experience with the radiosurgical treatment of AVM, it was deemed the ideal solution for certain cases because any area into the skull can be reached, because it is not necessary to open the brain, i.e., it is bloodless, and because its action is slow and progressive with no risk for the neighboring structures.

One of the most important conditions of radiosurgery is that the energy is delivered in a single dose.

Based on the principle that employing external beams of ionizing radiation, precise destruction or irradiation of deep-lying targets can be produced, we have developed a technically different system using a linear accelerator (Betti and Derechinsky 1981; 1983a; 1983b; 1984).

Methodological Considerations

Talairach's stereotactic frame is the nucleus of the reference system because it allows: a) a high precision fixing of the frame of the bone; b) all neuroradiological techniques employed are real-sized and stereoscopic in order to avoid radiological deformation (produced by the conicity of the X-ray beams); c) the possibility of a perfect repositioning of the frame as in the first diagnosis time, under identical geometrical conditions; d) some time in between to select the best strategy and to test different dosimetric computarized studies.

For diagnosis we used CT scan in stereotactic conditions with a plastic Talairach's frame. In the particular case of AVM, the angiography is taken in stereotactic conditions, real size and stereoscopy at 3 images per second. In some particular cases, ventriculography in stereotactic and stereoscopic conditions is added.

If our desire is to employ radiosurgery, precision in the spatial definition of the target volume is of the utmost importance.

Talairach's frame was modified keeping the identical geometric characteristics as those of the original one; this modification allows an easier repositioning of patients into it, and the possibility of placing the head in several and more confortable positions during the rotation of the frame.

The frame of our device makes it possible to place the center of a target volume in the isocenter of the linear accelerator. The frame can also rotate around a transverse axis. The combination of both movements: rotation of the accelerator and rotation of the stereotactic frame, covers a spherical sector. This spherical sector represents the surface of all possible portals of entry.

Using the pendular movement of the accelerator the whole dose is administered through infinite points in a 140° arc. This coronal diadem may cover a sagittal angle of 120°, spacing between them by some degrees in relation to the collimator diameter.

A special movable seat allows us to reduce body irradiation and provides more confort during treatments.

A collimating device for the 10 MeV beams of the linear accelerator, Clinac 18, Varian, was added in order to obtain a reduced penumbra. This secondary collimator is 80 cm long and works with 6, 7, 8, 10, 12, 14, 16 and 20 circular section diameters.

The Procedure

Some characteristics of our methodology are very important:

1. All types of neuroradiology are in stereotactic conditions: CT scan, angiography, ventriculography and serial biopsies.
2. Angiography and ventriculography are always of a stereoscopic kind and realsized in such a way that we can superimpose the pictures.
3. The time between diagnosis and therapy is very important to make all previsionnal computarized dosimetry as necessary and to plan with no possibility of failure the best strategy of treatment. In order to limit the injuries around the target volume we can choose an irradiation volume more similar to that of the nest whose shape can be spherical, fusiform or any other. In some cases we can add different volumes in order to cover up the volume of the AVM.
4. On the stereoscopic antero-posterior X-ray film we draw each projection of the planes or diadems to be used in order to obtain the most precise coronal contour. These contours are used to calculate the exact depth of the target volume for each point of the cranial convexity. Only the vertical plane is really coronal. The others have different obliquity in relation to the space between them. The computer must consider this angular function which modifies the distance between the isocenter and each point of the contour.
5. When the irradiation program is finished and the irradiation day is fixed, the patient is anesthetized with a slight neuroleptho-anesthesia, and the irradiation pins are introduced into the bone, the frame is put into place, the mechanical coordinates controlled and X-ray pictures in orthogonal positions are taken. If the superimposed films of the first diagnosis time and of the therapeutical time are in an identical position (only one image of all bone structures and metallic indicators for both films), we can fix the pins with a special device, and after that the frame is taken out. Slight anesthesia is interrupted and the patient is taken to the radiation room.

6. The Multibeam Convergent Irradiation Unit (UMIC) is brought under the accelerator and fixed to the floor with special anchors. The isocenter is checked with a mechanical system. The coordinates of the target volume are placed in the irradiation frame. The patient is placed into the irradiation frame and the pins fixed in our ordinary way, with the same depth in millimeters as that of the frame when it is placed again on the patient's skull. The pins retentive device is then taken out. The irradiation is carried out according to the plan. The patient is controlled by means of a T.V. camera. In order to avoid body irradiation, we can place the patient's body in positions out of the beams.

The lesion is referred to the mechanical coordinates, and so we define not only its volume and size but its exact spatial position.

On the treatment day we control the replacement of the stereotactic frame, and fix the pins in a non-movable way. The same spatial position of the target volume is placed in the irradiation frame and the target volume is made to coincide with the isocenter of irradiation and rotation of the linear accelerator. This simple transposition of the AVM nidus from the X-ray picture to the accelerator isocenter in a very precise way is the key of our method.

The position of the body during irradiation is another important point because the seated position is the only one permitting the choice of diadems of irradiation at the cranial level without irradiation of other parts. The body displacement behind the projection of the irradiation beam during the irradiation of the most posterior cranial diadems or the forward displacement when the frontal diadems are concerned, allows us to give only an exiguous radiation dose. Other positions such as the dorsal decubitus that we have used in our first trials result in important doses.

UMIC: Dosimetrical Aspects

In the UMIC technique as in conventional treatments, beam behavior must be considered to obtain successful relative dose distributions and in the assessment of accelerator parameters, to perform the irradiation strategy. Then a theorical simulation of treatment must be made, in which two models interact: the beam model and the patient model. Here follows a description of both models and dose computation.

The Beam Model

A Varian Clinac 18 Linear Accelerator is used to deliver the target dose. The "standard" beam, X-rays of 10 MeV of nominal energy, is over collimated to obtain narrow circular section beams from 8–20 mm or a rectangular field of 7×21 mm, in both cases, at the isocenter. Special brass pieces are added to gantry's head to define such fields. The beam model generation involves two steps. An experimental one, in which relative and absolute radiation parameters are measured and a theoretical one, in which a beam model is selected to represent experimental results.

Experimental Measurements

Dosimetric measurements were carried out with conventional methods, thermoluminiscent dosimeters and ion chamber. TLD of LiF and CaF of 3 × 3 × 1 mm were used (Thomasz et al. 1983); readings were performed with a Harshaw-200 reader, and each dosimeter was previously heated for 1 h at 400°C and for 3 h at 100°C, and individually calibrated. Readings after irradiation were done at a maximum temperature of 300°C and a heating speed of 10°C/s with individual reading corrections. Two different TLD dosimeters were used to analyze energy perturbation generated by the over-collimation. A PTW 30-329 ion chamber assembled to a reader Victoreen RADCOM III was selected for its little volume of collection and beam perturbation. Techniques of film dosimetry are now used to add extra information.

Two kinds of measurements were performed: in- and off-axis of the irradiation beam.

Axis Measurements

Experimental data for the Output Factor (Cis), defined as the dose per monitor unit at 5 cm depth and 95 cm source-surface distance, and Tissue-Maximum-Ratio (TMR) were obtained. Measurements were performed in a lucite water phantom ($\delta = 1.19$ g/cm^3), and water-lucite equivalence assumed as in the AAPM TG'21 protocol. For the smallest beam calibration, only TLD was carried out; for greater beams, TLD and ion chamber were used.

Off-Axis Measurements

Off-Axis-Ratios (OAR) measurements were indirectly carried out with TLD detectors distributed in three orthogonal axis of a semispheric 7 cm radius water phantom with lucite walls. Single and multiple irradiation arcs were used to obtain a smaller dose gradient in the penumbra zone to avoid poor resolution since TLD size (3 × 3 × 1 mm) was not small enough. A computer program makes the deconvolution and generates a single beam off-axis-curve.

Rando Measurements

The simulation of therapeutical practice was carried out with a Rando equivalent tissue phantom. TLD-700 lithium floride dosimeters were distributed both deeply and on the surface to assess mean absorbed dose for several organs and tissues (Thomasz et al. 1983).

Theoretical Considerations

A beam model was selected to reproduce experimental data. In agreement with magnitudes showed in Fig. 1, Percentage-Depth-Dose (PDD) for a point of (u_1, u_2, u_3) coordinate is given by:

$$\mathrm{PDD}\,(u_1, u_2, u_3) = P_0\,(u_1)\,\mathrm{OAR}\,(u_1, u_2, u_3) \qquad (1)$$

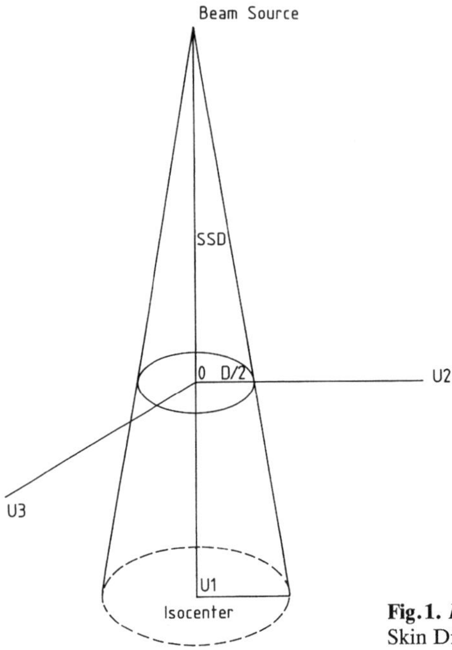

Fig. 1. *D/2:* Beam Radius at the Surface. – *SSD:* Source Skin Distance – *UMIC:* Beam Coordinate System

where, $P_0(u_1)$, the beam profile, is given by:

$$P_0(u_1) = \left(\frac{SSD + u_{1M}}{SSD + u_1}\right)^2 \left[A \exp(-m(u_1 - u_{1M})) - (A - 100) \exp(-g(u_1 - u_{1M}))\right] \tag{2}$$

with

$$A = 100 + A_1(1 - \exp(-A_3L)) + A_2 1L \exp(-A_3L) \tag{3}$$

$$m = B_1(1 - \exp(-B_3L)) + B_2 \exp(-B_3L) + B_4L \exp(-B_3L) \tag{4}$$

u_{1M} = Depth of maximum dose for a single beam

$$L = \frac{2 L_1 L_2}{(L_1 + L_2)} \text{ for real rectangular fields } L_1 \times L_2 \text{ at the surface} \tag{5}$$

and

$$L = \sqrt{\pi \frac{D}{2}} \text{ for real circular fields of D diameter at the surface} \tag{6}$$

For OAR (u_1, u_2, u_3) two functions were selected. Defining u as:

$$u = \left(\frac{SSD}{SSD + u_1}\right) \frac{2}{D} \sqrt{u_2^2 + u_3^2} \qquad \text{for circular fields} \tag{7}$$

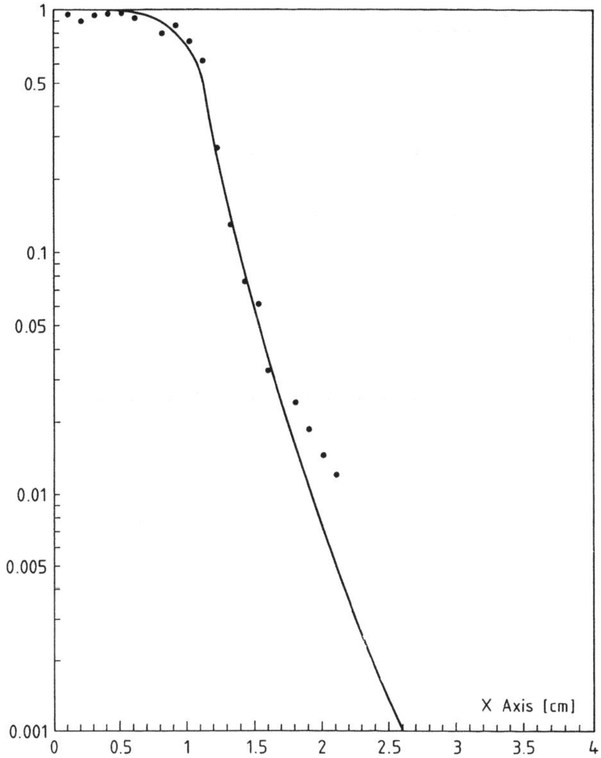

Fig. 2. UMIC-Relative Dose Distribution – X axis Rectangular Collimator. – Calculated Curve · Experimental measurements

and

$$u = \left(\frac{SSD}{SSD + u_1}\right) \sqrt{\left(\frac{u_2}{L_1}\right)^2 + \left(\frac{u_3}{L_2}\right)^2} \qquad \text{for rectangular fields,} \tag{8}$$

then

$$OAR\,(u_1, u_2) = \begin{cases} 1 - \int_{-\infty}^{u} \frac{1}{\sqrt{2\pi\sigma^2}} \exp\left(-\frac{1}{2}\left(\frac{u-1}{\sigma}\right)^2\right) du \text{ for } u \leqslant 1 \\ 0.5 * \exp\,(C_1\,(u-1)^{C_2}) \text{ for } u > 1 \end{cases} \tag{9}$$

Two different functions were selected for OAR, since "penumbra" factors are modified by the overcollimation. Parameters like:

$A_i\ (1 \leqslant i \leqslant 3)$, $B_i\ (1 \leqslant i \leqslant 4)$, $C_i\ (1 \leqslant i \leqslant 2)$, G

were calculated using a least square fit technique, minimizing the s function given by:

$$S = \sum_i (y_i - P_i)^2 \tag{10}$$

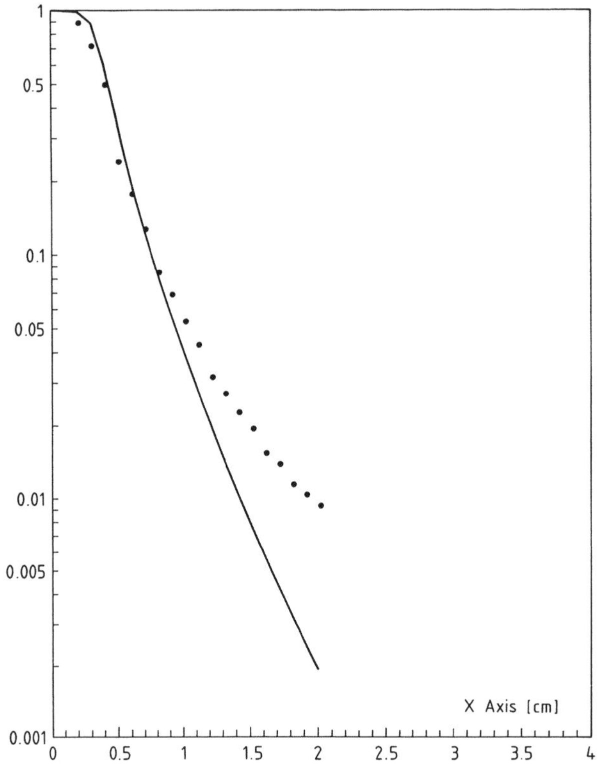

Fig. 3. UMIC-Relative Dose Distribution – Z axis Rectangular Collimator. – Calculated Curve · Experimental measurements

solving a non linear equation system, in which y_i are experimental data and P_i calculated values.

P_0 parameters were first fitted using the TMR measurements considering the relational expression

$$TMR\,(u_1, L_{u_1}) = \left(\frac{SSD + u_1}{SSD + u_{1M}}\right)^2 \frac{1}{100}\; P_0\,(u_1, L, SSD) \tag{11}$$

with

$$Lu_1 = \left(\frac{SSD + u_1}{SSD}\right) L \tag{12}$$

Then, this P_0 values were used to fit OAR parameters. Figs. 2 and 3 show fit accuracy for a rectangular field. This pictures were selected to represent both calculated responses simultaneously.

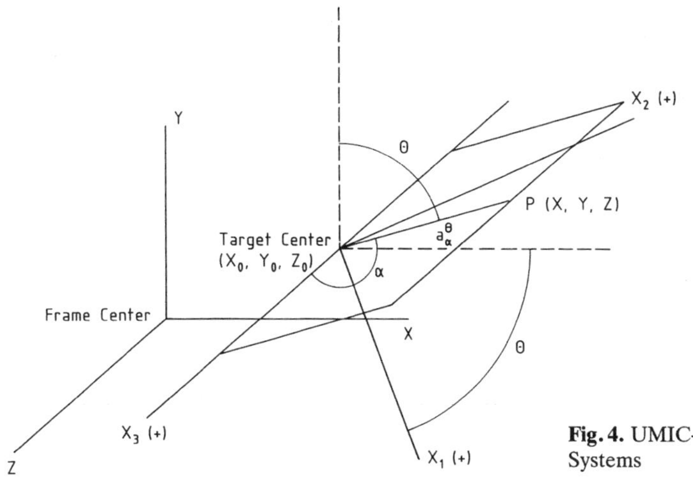

Fig. 4. UMIC-Associated Coordinate Systems

The Patient Model

Patient model is related directly to Talairach's stereotactic frame. A three dimensional cartesian coordinate system related to it is shown in Fig. 4.

Then, any point in the brain is represented by its coordinates (X, Y, Z). When the irradiation volume is defined, one or more targets can be selected, each of them with its center coordinates (X_0, Y_0, Z_0) in the Talairach's system. In the UMIC technique a target dose is delivered by successive irradiation arcs according to different planes, rotated a θ angle to the vertical position. If coordinate systems $(X_1^\theta, Y_1^\theta, Z_1^\theta)$ centered in the target and rotated a θ angle are connected to each plane, all of them have a common axis, X_3^θ (Fig. 4). The patient model is generated as a set of distances between the target center and the patient's contour according to each cross-fire direction into each plane of irradiation. A vertical projection of patient's contour is depicted for all irradiation planes from the lateral and antero-posterior films obtained in stereotactic and stereoscopic conditions. According to Fig. 4, defining a_α^θ as the distance between target center and patient's contour in a α direction to X_3 axis to plane, and r_μ^θ and μ the corresponding magnitudes in the vertical projection, we have:

$$\begin{aligned} a_\alpha^\theta &= r_\mu^\theta \left(\cos^2\alpha + \cos^2\theta \sin^2\alpha\right)^{-1/2} \\ \mu &= \tan^{-1}(\cos\theta \tan\alpha) \end{aligned} \tag{13}$$

Then, selecting the appropriate μ direction for each X position, measurements of r_μ^θ can be done and a_α^θ so obtained.

Dose Computation

In the irradiation strategy selection, radiosensitivity of tissues surrounding the target volume must be considered. Thus, relative dose distribution is the main aspect to be considered. Computer programs (Galmarini and Alvarez 1984) for a Hewlett Packard 150 personal computer were developed in order to calculate both dose distributions and accelerator parameters, like number of monitor units and monitor units/degree of rotation.

Relative Dose Distributions

The irradiation method, in detail, consists of several moving fields irradiations by one or more arcs included in the same plane, for a defined number of planes for each target. In order to obtain the best relative dose distribution, target positions, plane orientations and angular arc apertures can be selected and externally weighted. Defining as

w_i^t external weight of the i-target $(1 \leqslant i \leqslant t)$

w_{ij}^p external weight of plane j of i-target $(1 \leqslant j \leqslant p(t))$

w_{ijk}^s external weight of sector k of plane j of target i $(1 \leqslant k \leqslant s(p))$

and m_k the number of directions considered to perform calculations defined as

$$m_k = \frac{A_k}{q} \tag{14}$$

where A_k is the angular aperture of the k-sector and q is the angular interval between two directions. Thus, the Percentage Depth Dose for any point (X, Y, Z) in the associated Talairach's system is given by:

$$\mathrm{PDD}\,(X, Y, Z) = 100 \tag{15}$$

$$\frac{\sum_{i=1}^{t} \sum_{j=1}^{pt} \sum_{k=1}^{sp} \sum_{l=1}^{m_k} w_i^t\, w_{ij}^p\, w_{ijk}^s\, \mathrm{PDD}_{ijkl}\, [u_1 (X, Y, Z), u_2 (X, Y, Z), u_3 (X, Y, Z)]}{\sum_{i=1}^{t} \sum_{j=1}^{pt} \sum_{k=1}^{sp} \sum_{l=1}^{m_k} w_i^t\, w_{ij}^p\, w_{ijk}^s\, \mathrm{PDD}_{ijkl}\, [u_1 (X_N, Y_N, Z_N), u_2 (X_N, Y_N, Z_N), u_3 (X_N, Y_N, Z_N)]}$$

where (X_N, Y_N, Z_N) are the coordinates of the normalization point. Functions u_1, u_2, u_3 are given by

$$\begin{aligned} u_1 &= a_\alpha^\theta - \sqrt{(X_2^\theta)^2 + (X_3^\theta)^2}\ \cos\left[\tan^{-1}\left(\frac{X_2^\theta}{X_3^\theta}\right) - X\right] \\ u_2 &= \sqrt{(X_2^\theta)^2 + (X_3^\theta)^2}\ \sin\left[\tan^{-1}\left(\frac{X_2^\theta}{X_3^\theta}\right) - X\right] \\ u_3 &= X_1^\theta \end{aligned} \tag{16}$$

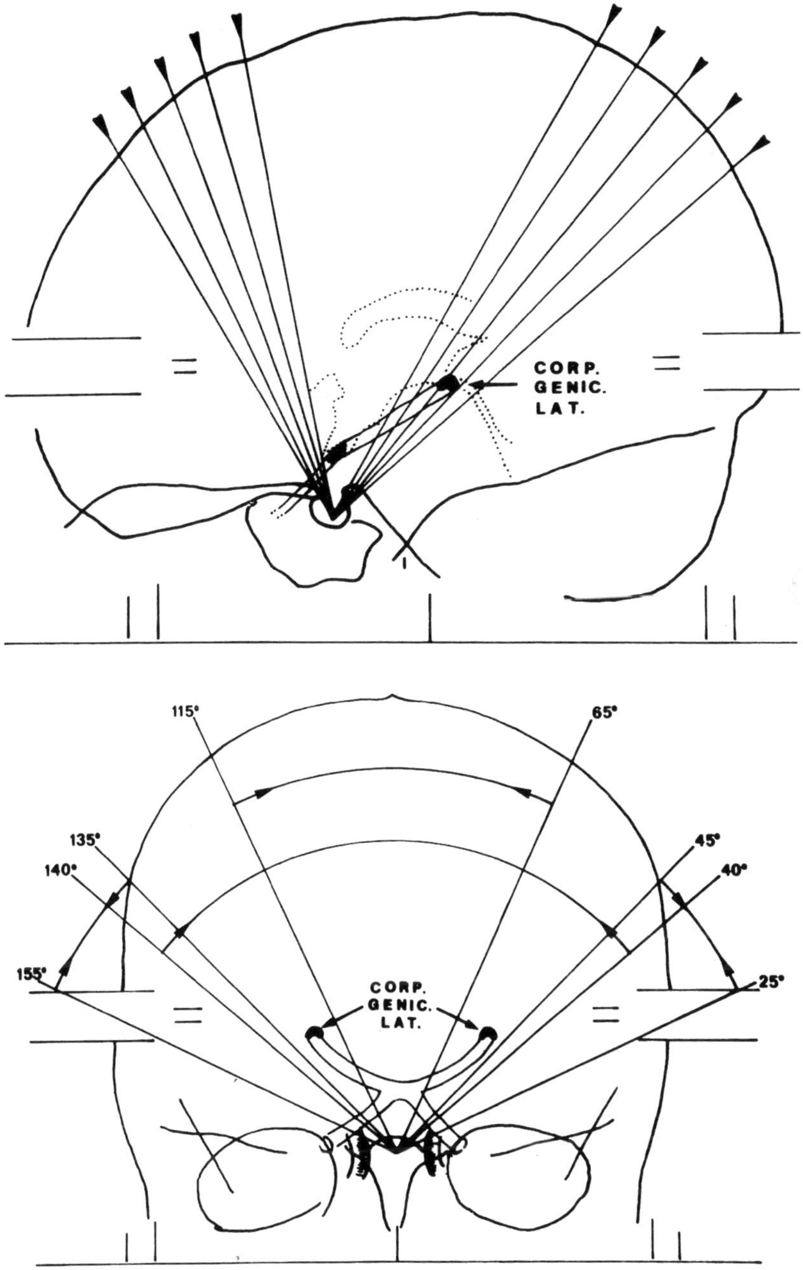

Fig. 5. Example of hypophysis irradiation avoiding all the optic pathway

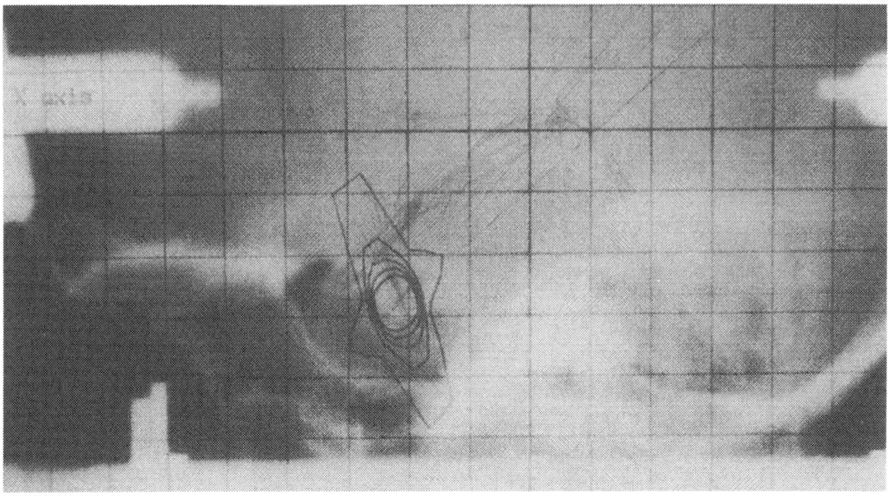

Fig. 6. Dosimetric results of the figure 5 projected over the patient's film

with

$$X_1^\theta = (X - X_{0_i}) \cos\theta - (Y - Y_0) \sin\theta$$
$$X_2 = (X - X_0) \sin\theta + (Y - Y_0) \cos\theta \qquad (17)$$
$$X_3 = Z - Z_0$$

where α is the direction of the l beam in the θ plane. Examples of calculated distributions are shown in Figs. 5 and 6. When it was possible, a 70–75% isodose level was selected to round the critical volume limits.

Accelerator Parameters Calculation

When relative dose distribution is selected, geometric parameters are recorded to be used to calculate the number of monitor units needed to perform the treatment. In spite of external weights, plane doses are balanced proportionally to corresponding average TMR to assess equal maximum doses for all planes. Using the same definition as before, and defining for each target

- D: dose at the target center
- D_j: dose delivered by j-plane
- D_{jk}: dose delivered by the k-sector of j-plane
- D_{jkl}: dose delivered by 1-direction in the k-sector of j-plane

we have,

$$D_j = \frac{w_j^p <TMR>_j A_j}{\sum_j w_j^p <TMR>_j A_j} D \qquad (18)$$

$$D_{jk} = \frac{w_{sk} \langle TMR \rangle_{jk} A_{jk}}{\sum_k w_{sk} \langle TMR \rangle_{jk} A_{jk}} D \tag{19}$$

$$D_{jkl} = \frac{TMR_{jkl}}{\sum_l TMR_{jkl}} D_{jk} \tag{20}$$

with

$\langle TMR \rangle_j$: average value of TMR for j-plane

$\langle TMR \rangle_{jk}$: average value of TMR for k-sector of j-plane

$\langle TMR \rangle_{jkl}$: TMR value for l-direction of k-sector of j-plane

A_j: angular aperture for j-plane

A_{jk}: angular aperture for j-sector of k-plane

Defining U_{jkl} as the number of monitor units for a special direction, it is calculated as

$$U_{jkl} = \frac{D_{jkl}}{\text{Cis } TMR_{jkl}} \tag{21}$$

Values for U_{jk} and U_j are given by

$$U_{jk} = \sum_l U_{jkl} \tag{22}$$

$$U_j = \sum_k U_{jk} \tag{23}$$

The number of monitor units by degree are defined as

$$U_{jk} = \frac{U_{jk}}{A_{jk}} \tag{24}$$

An option of fixed portals of entry can be used at any time in the treatment if something occurs avoiding arc therapy mode without any harm to the treatment quality.

Clinical Findings

During the last 30 months, between September 1982 and April 1985, we have treated 55 patients, all of them sent by other neurosurgical teams, at Antartida Hospital in Buenos Aires, who were part of the 110 patients irradiated with the Multibeam Convergent unit.

Table 1. Arterial supply to AVM

	Left	Right
Anterior cerebral artery	3	1
Pericallosal artery	1	3
Frontopolar artery		1
Recurrent artery of Huebner	1	1
Middle cerebral artery	1	4
Lenticulo striate arteries	2	4
Occipital artery	1	
Opercular arteries	1	2
Temporal arteries	2	
Parietal arteries	1	
Rolandic arteries	3	1
Anterior choroidal artery	4	4
Posterior cerebral artery	6	8
Parietal arteries		1
Temporal arteries	1	
Thalamo-perforating arteries	1	2
Thalamo geniculate arteries	1	
Superior cerebellar artery	1	3
Peduncular artery	2	

The size of an AVM may be small or large considering a 3 cm diameter as a reference and taking into account the maximum diameter for classifying them (Pelletieri 1980). We must emphasize that our films have only a 4% magnification and all patients were studied in three-dimensional angiography with 3 pictures taken per second.

We judge an AVM according to its being or not within the silent or nonsilent areas taking into account the clinical signs. The possibility of considering the AVM as cortical or subcortical is not easy to define, for some of them exhibit both characteristics; we classify AVM as midline ones considering the sagittal plane from the midline of the skull to 1 cm at both sides, and as superficially or deeply situated as defined by Pelletieri et al. (1980), limited by a frontal projection curve line from the sagittal sinus through the internal capsule to the base of the skull.

All patients are controlled by means of the CT scan every 6 months until an important reduction or modification in the size of AVM is seen. At that moment an angiography is taken to show the real changes in the malformation vessels.

The latency between the multibeam treatment and the obliteration of AVM is not standard and in certain cases we have had the evidence of disparition of the AVM in 9 months or even of its reduction in 6 months, but in other cases it takes from 18–30 months.

Irradiation with our technique apparently gives the same benefit as the proton-beam irradiation, because no rebleeding was confirmed in our patients remembering

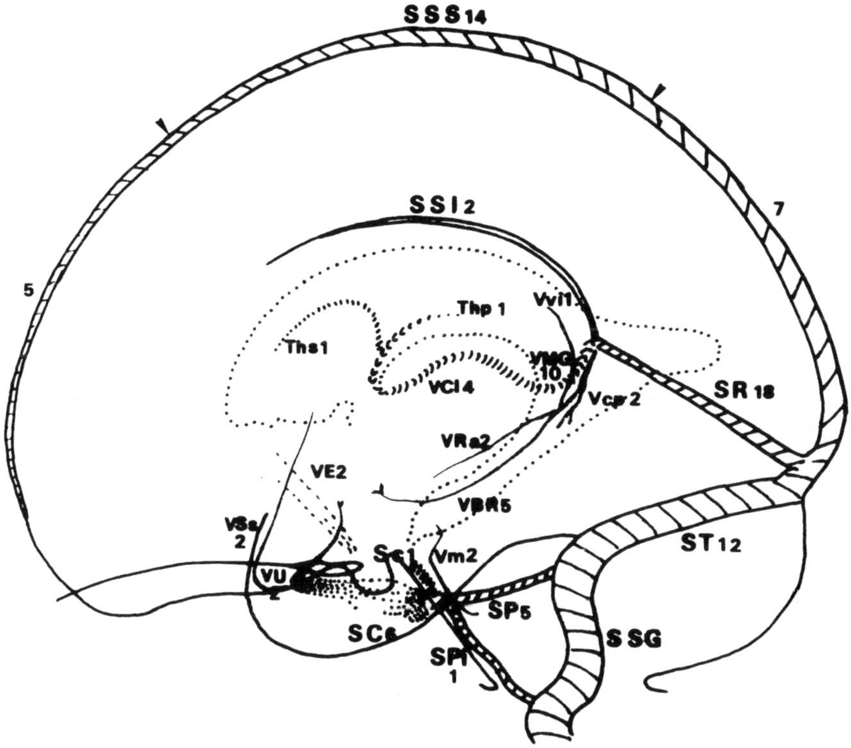

Fig. 7. Frequency of AVM drainage. *SSS:* sinus sagittales superior; *SSI:* sinus sagittales inferior; *SR:* sinus rectus; *ST:* sinus transversus; *SP:* sinus petrosus; *SSg:* sinus sigmoideus; *SC:* sinus cavernosus; *VGM:* vena magna Galeni; *VBR:* vena basalis Rosenthali; *VCI:* vena cerebri interna; *Ths:* vena thalamica superior; *Thp:* vena thalamica posterior

that the hemorrhage rate in large series was an annual rate of 2%–3%. The rebleeding rate is 6% in the first year after hemorrhage but about 10% of the patients die with the first hemorrhage and 20% with each rebleeding, and the 60% of survivors are symptoms free (Graf et al. 1983) but nobody knows who will be the survivors.

Perret and Nishioka (1966) stated 27 recurrences of hemorrhage in 103 non-operated patients, and Foerster et al. (1972) reported 8%–9% of rebleeding of their non-operated patients. Nystrom (1975) found 15% of one recurrence hemorrhage and 9% had had two or more recurrences.

Pregnancy increases the risk of hemorrhage in patients with AVM (Robinson et al. 1974) from the normal 10%–87%. This is the reason why we treated using our particular technique, our case no. 41, a pregnant woman with 2 bleedings at the 2nd and 5th and a half months of pregnancy, and though in the second episode she was in a coma with hemorrhagic ventricules, she was able to recover successfully after that.

No patient bleeds after irradiation (Kjellberg et al. 1983). One patient died three months after treatment from an obstruction of the derivative valve.

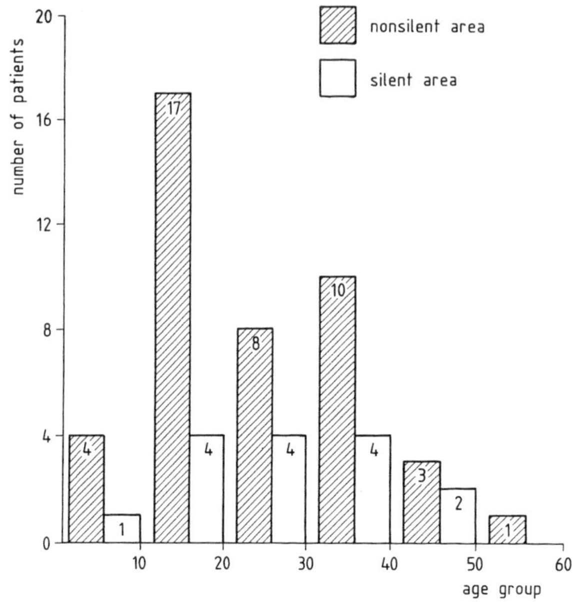

Fig. 8. Distribution of AVM in *"silent"* and *"nonsilent"* areas vs. age

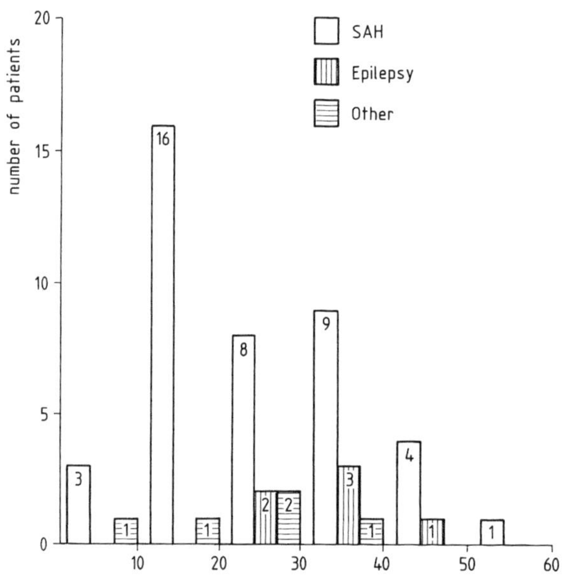

Fig. 9. Distribution of various diagnosis at admission vs. age

The pregnant woman treated during the 5 months of pregnancy, has given birth to a healthy boy by means of a cesarean section.

We must emphasize that all our patients were previously considered non-surgical or high-risk cases by other neurosurgical teams and sent to us after rejecting the possibility of a direct surgical approach.

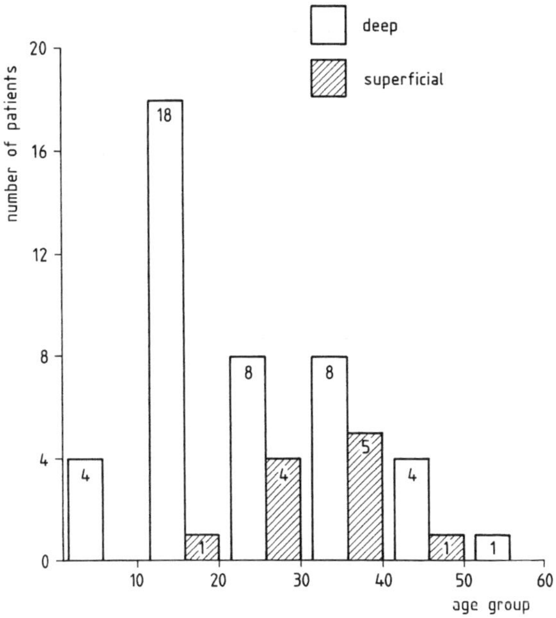

Fig. 10. Distribution of deep and superficial AVM vs. age

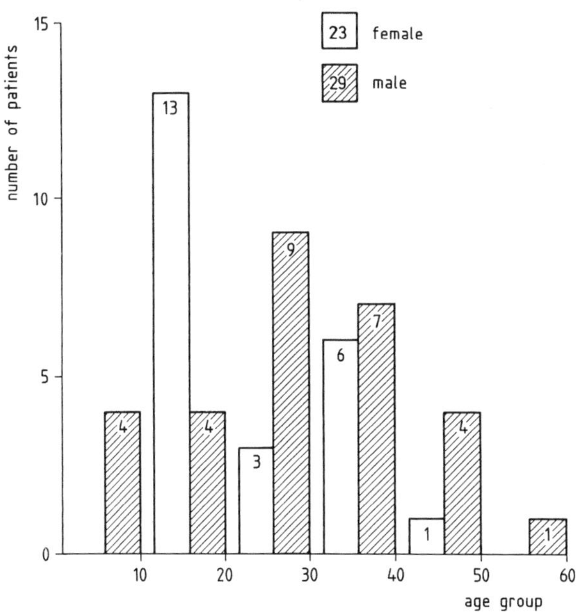

Fig. 11. Distribution of male and female patients vs. age

The treatment of 2 separated AVM in the same patient and in only one session, with occlusion of both is another proof of success in the usage of this technique.

These are our first results and we are activating the follow-up controls in order to determine the exact number of patients with AVM occlusion. The time elapsed is

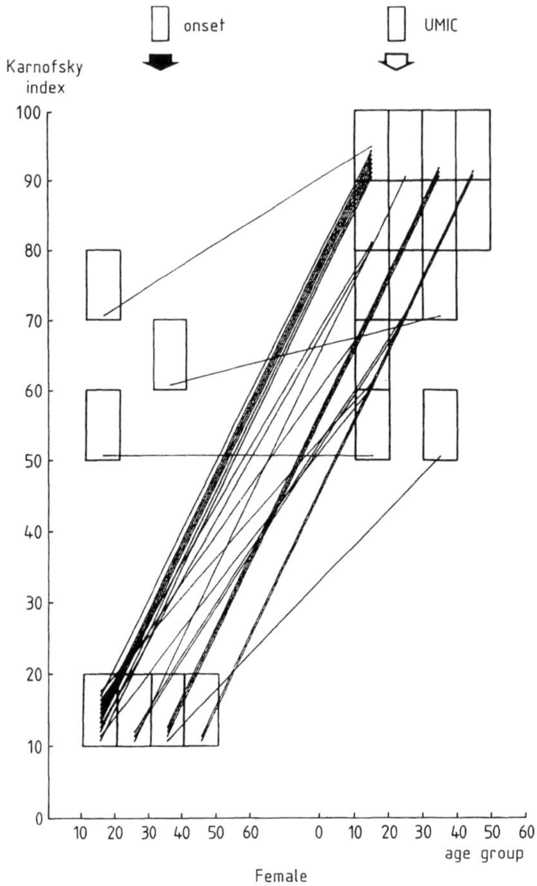

Fig. 12. Karnofsky index of female age group at onset and at the moment of UMIC treatment

very different in the cases presented, and we can suppose that a great number of the remaining cases will result in the occlusion of their AVM.

These cases have 17 months as average occlusion time and represent the 34% of the cases in that period. Other cases are controlled by CT scan and we demand the angiography only when we find modifications of CT. Some patients, due to distance between their dwelling place and the hospital, send us their controls with a certain delay.

We feel that our method is the choice treatment in non-surgical cases and an alternative possibility in cases of high risk because we have no mortality and our morbility of 2% may be reduced even more.

As we have seen, the ideal cases are the AVM in deep areas, of small size and particularly in women.

If small AVM are easy to treat we can also see that large lesions may be irradiated with similar success, if we can divide them into portions and apply the treatment 2 or 3 times.

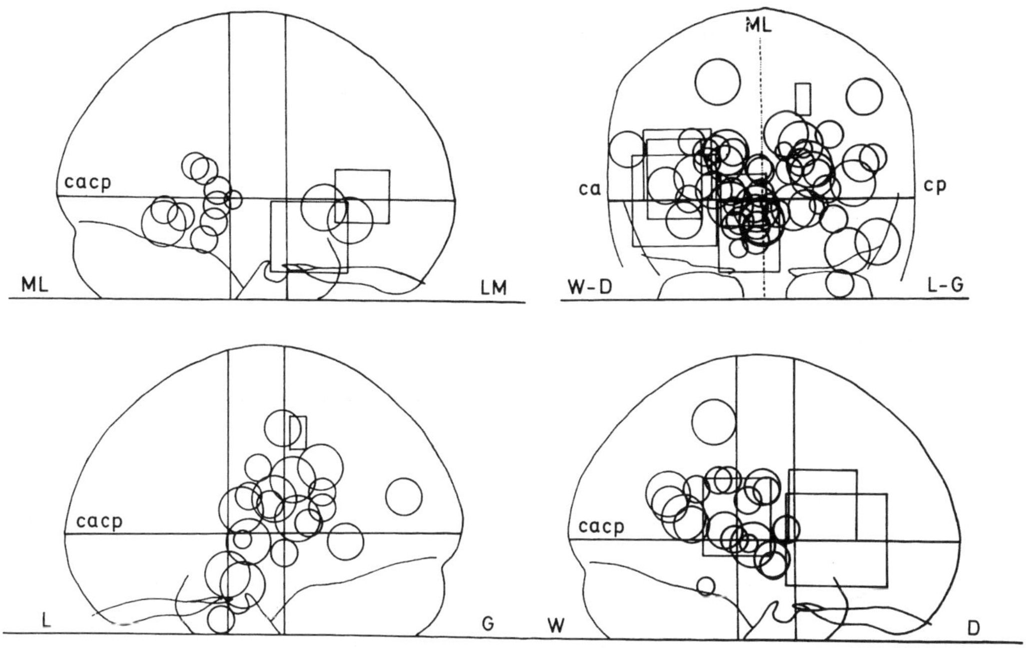

Fig. 13. Projection of AVM according to the collimators employed in the irradiation

In the study of our cases we have only considered the patients treated before December 1984 (44 cases). We cannot demonstrate any evident relationship between the size of the AVM and the time when the volume reduction started.

In most of the cases, the AVM disappeared in the CT suddenly. In some cases the last CT showed slight modifications in contrast or in the size. The disparition time varies from 1–2 years. We have controlled 13 cases during 2 years or more. If we consider the short time elapsed between treatment and results (AVM fading away), we cannot evaluate the real efficacy of the method notwithstanding the reduction in size in 3 cases and the disappearance in 8 cases in CT and control angiography are the evidence of a positive action of the UMIC irradiation.

The modalities of disappearance are completely different as in the case 26 that showed litte modifications during 2 years and that vanished suddenly, and others that reduced progressively (Figs. 7–15).

Discussion

Prof. C. G. Drake (1983) gives 5 options for the management of cerebral arterio-venous malformations: 1) leaving them alone; 2) surgical excision; 3) embolization; 4) high-energy therapy (photons, gamma-rays or X-rays); and 5) combinations of options 2, 3 and 4.

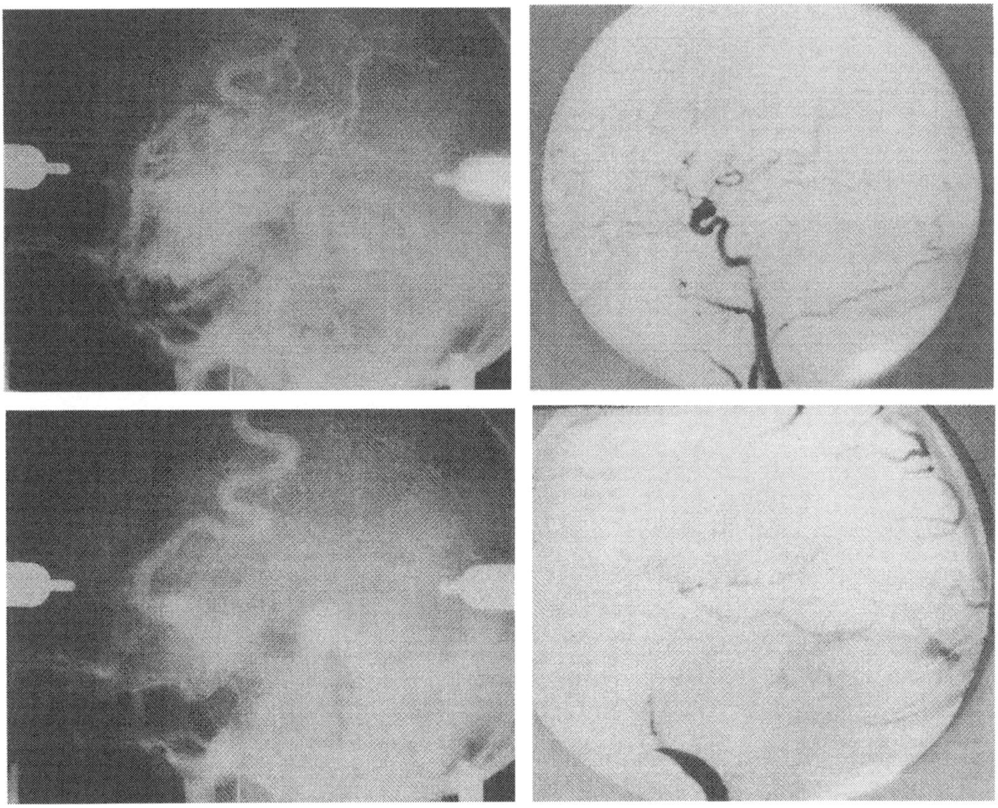

Fig. 14. Case no. 26: M.A.T., male, 38 years old, with a great AVM on the *left side* treated with UMIC, and on the *right* side the response 13 months later with total disappearance of the malformation

It is generally accepted that option 1 is the last because the need of giving some solution in symptomatic cases obliges us to consider the other 4 possibilities.

The resection with the surgical microscope has proved effectual in widening the bounderies including lesions beside or even in vital cerebral regions. It should be stressed that this assessment is meant only for qualified neurosurgeons with refined techniques and, we add, with an important technical background and great experience. It is not the case for wizardry beginners.

Option 3 is a special case, and with the development of plastics the possibility of introducing catheters becomes more common every day.

But option 4, in our experience, has been enriched with option 5. In 1969, Johnson emphasized the value of radiotherapy in cases of inoperable AVM.

Encouraged by this statement, in 1970 Steiner (Steiner et al. 1972) began the treatment of AVM keeping in mind that blood vessels are easily injured by ionizing radiation. He irradiated different targets such as the feeding vessels and the nest because the smaller they are, the more sensitive they become. Single doses of 50 Gy or

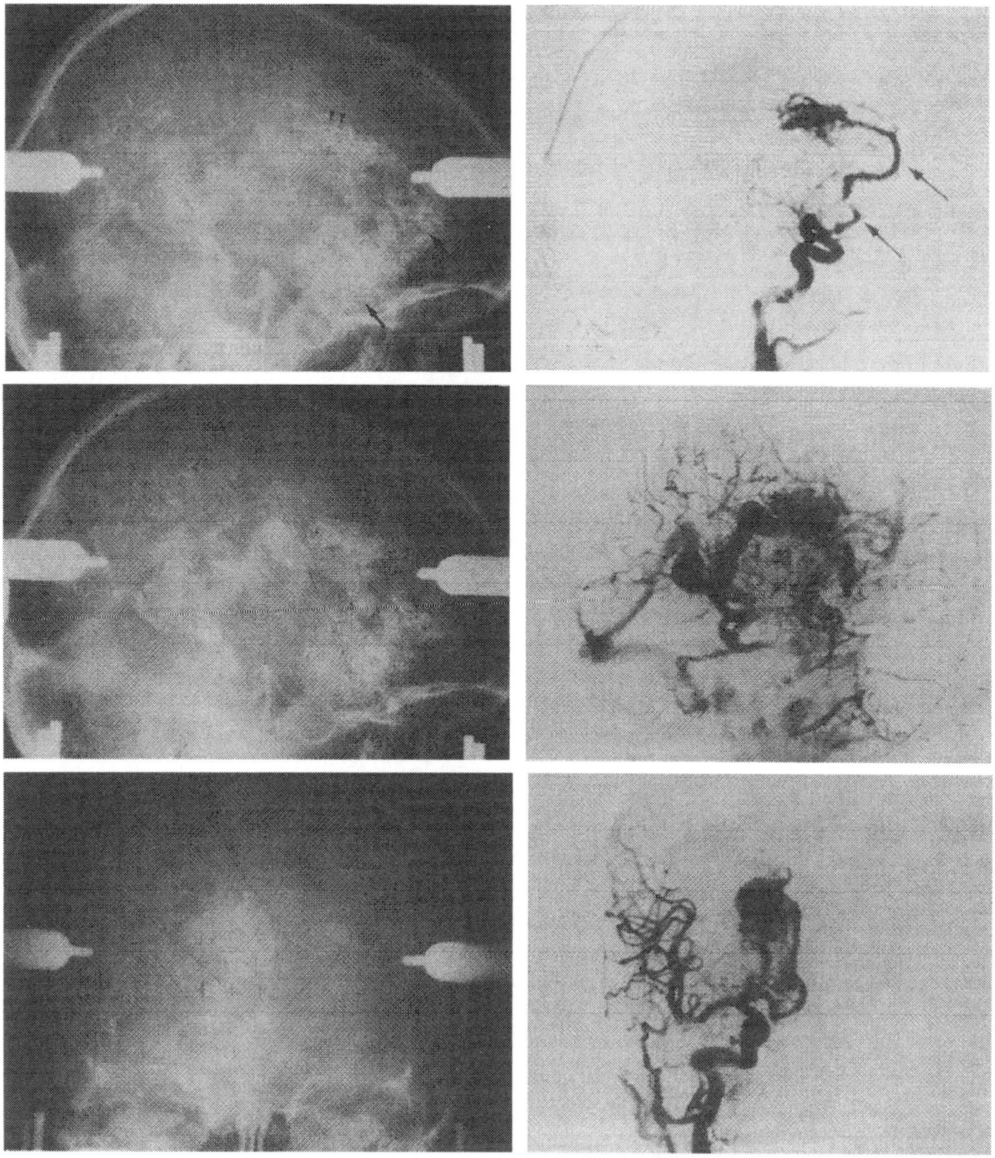

Fig. 15. Case no. 12: F. de L., female, 33 years old. On the *left,* extensive malformation partially irradiated (only the anterior portion). 21 months later, on the *right,* the results show the disappearance of the irradiated part of the AVM

more were used and after a period varying from 1–2 years the 85% of malformations had disappeared and the vasculature in the region was normal (Steiner et al. 1979).

Our experience began in 1983 and so far we have used the Multibeam Convergent Irradiation Unit (UMIC) in the treatment of 110 cases, 55 of which were patients with AVM.

The nidus of the AVM was treated with our technique using the most appropriate collimator diameter in each case. The reduced penumbra in our device allows us to treat lesions that lie very close to dangerous areas, such as the pyramidal tract, the optic pathway, the basal ganglia, the motor cortex, the speech cortical areas, the hypothalamus and the brain stem.

When an absolute dose is prescribed, the subsequent problem is to select the relative dose to surround the surface; in other words, the smallest dose in the target. Backlund (1979) describes a spatial agreement between the target surface and a 50% isodose level. We decided to use 70%–75% isodose level and a smaller absolute dose; in this way we obtain a more homogenous distribution with the same dose in points which are out of the target. Our results seem to be apparently the same as those of the Gamma Unit's, but ours have been obtained using a smaller target dose.

Conclusion

The possibility of treating patients with the multibeam system is thus demonstrated and our little contribution to this field gives further support to this form of treatment.

Most of our patients were considered inoperable cases by other neurosurgeons, and even the family or the patient himself deemed the risk of the procedure very high in several opportunities.

The concept of risk adds a new dimension to the character of operability which only means that the lesion "is technically amenable for operation".

Some variables such as superficial or deep location and size, are used as a valid argument in assessing operability, but the risk of rebleeding is also important.

In our groups the size of AVM is related to a high incidence of hemorrhage; another important factor is sex.

The occlusion of AVM treated with Multibeam Convergent Irradiation Unit with a Linear Accelerator shows that it is a reliable instrument for treating different types of malformations.

In fact, no rebleeding has occurred in cases treated with this method up to the present time. We can assume that the hyperselective irradiation with UMIC offers certain protection against hemorrhage.

We have had 2 complications, only one of which was an AVM, and both originated in technical defects avoided after that.

No mortality as a consequence of treatment or reccurrence of hemorrhage occurred during the 30 months after the first patient was treated.

We treated not only small and deep lesions but large AVM ones, by portions and at different times. As the results were successful, we extended the applications of the multibeam technique.

We use a computer programme for calculating doses in order to avoid complication. We employ composed target volumes of irradiation in order to obtain the most successful results with the minimum dose and the lowest possible risk.

"In the absence of an absolute rule to guide his decision, the surgeon must carefully plot each borderline case against the abscissa of danger and the ordinate of benefit".

H. W. Pia, Giessen Symposium 1974)

References

Backlund EO (1979) Stereotactic radiosurgery in intracranial tumours and vascular malformations. In: Kranenbühl H, Brihane J, Loew F, Logué V, Mingrino S, Pertuiset B, Symon L, Troupp H, Yasargil MG (eds) Advances and technical standards in Neurosurgery, vol 6. Springer, Berlin Heidelberg New York Tokyo, p 6

Betti OO, Derechinsky VE (1981) Irradiation stéréotaxique multifaisceaux. Comm Soc Neurochir Langue Franç, Paris, 30.11.1981

Betti OO, Derechinsky VE (1983a) Irradiación hiperselectiva encefálica con acelerador lineal: "Unidad Multihaz Jean Talairach". Prense Médica Argent, vol 70/3: 102–107

Betti OO, Derechinsky VE (1983b) Irradiation stéréotaxique multifaisceaux. Neurochir 29: 295–298

Betti OO, Derechinsky VE (1984) Hyperselective encephalic irradiation with linear accelerator. Acta Neurochirurgica [Suppl] 33: 385–390

Drake CG (1983) Arteriovenous malformations of the brain: The options for management. N Engl J Med 5: 308–310

Foerster MD, Steiner L, Häkanson S (1972) Arteriovenous malformations of the brain. A longterm clinical study. J Neurosurg 37: 562

Galmarini DH, Alvarez AG (1984) Planificación en radioterapia externa con microcomputadoras. IV Congreso CRILA, 4–8 abril 1984, Buenos Aires, Argentina

Graf CL, Perret GE, Torner JC (1983) Bleeding from cerebral arteriovenous malformations as part of their natural history. J Neurosurg, 58: 331–337

Johns HE, Cunningham JR (1983) The Physics of Radiology (4th edn). Thomas, Springfield, Illinois

Kjellberg RN, Hanamura T, Davis KR, Lyons SL, Adams RD (1983) Bragg-Peak proton beam therapy for arteriovenous malformations of the brain. N Engl J Med, 5: 269–274

Nyström SH (1973) Congenital arteriovenous malformations of the brain. Helsinki, Finland (quoted by Pelletieri)

Pelletieri L (1980) Surgical versus conservative treatment of intracranial arteriovenous malformations. Acta Neurochir Suppl 29: 86

Perret GE, Nishioka H (1966) Report on the cooperative study of intracranial aneurysms and subarachnoid hemorrhage. Arteriovenous malformations. J Neurosurg 25: 467

Robinson JL, Hall CS, Sedzimir CB (1974) Arteriovenous malformations. Aneurysms and pregnancy. J Neurosurg 41: 63

Steiner L, Leksell L, Greitz T, Foerster DM, Backlund EO (1972) Stereotaxic radiosurgery for cerebral arteriovenous malformations. Report of a case. Acta Chir Scand 138: 459–464

Steiner L, Greitz T, Backlund EO, Leksell L, Noren G, Rähn T (1979) Radiosurgery in arteriovenous malformations of the brain. Undue effects. In: Szikla G (ed) Stereotactic cerebral irradiation, INSERM Symp 12. Elsevier, North Holland, pp 257–269

Thomasz E, Spano F, Massera G (1983) Determinación de las dosis absorbidas en irradiaciones de Base de Cráneo con Rayos X de 9,1 MeV. IV Congreso CRILA, 4–8 abril, 1983, Buenos Aires, Argentina

Linear Accelerator Radiosurgery: Technical Note

F. Colombo, A. Benedetti, F. Pozza, A. Zanardo, R. C. Avanzo, G. Chierego, and C. Marchetti

Introduction

Single-dose, stereotactically focalized, external irradiation (radiosurgery) has gained wide acceptance in neurosurgery as a useful and safe technique for dealing with neurological diseases, intracranial tumors, and arteriovenous malformations (Leksell 1971, Kjellberg et al. 1972). Nevertheless, the application of this technique was until very recent times restricted to the Karolinska Institut of Stockholm, owing to the high costs of the Gamma unit and relative lay out. Proton beam radiosurgery also requires complicated and expensive equipment and cannot be made widely available.

At the Neurosurgical Department of Vicenza we have developed a new radiosurgical technique which utilizes a common isocentric linear accelerator coupled with our stereotactic apparatus which we have employed in clinical practice since 1982 (Colombo et al. 1983, 1985). By this straightforward technique any isocentric linear accelerator can be converted into radiosurgery apparatus. The extra equipment needed is neither complicated nor expensive.

Materials and Methods

The Working Principle

The technique is based on the composition of multiple concentric arc irradiations focalized in the stereotactic target. The selected target held at the rotational isocenter of the linear accelerator. A first arc irradiation is performed. The patient is then rotated around a vertical axis passing through the target and arc irradiations are repeated on different angular approaches, spread over the convexity of the skull.

Physical and Dosimetric Evaluation

Before clinical trial, the performance of our technique in terms of safe dose delivery, dose concentration, and toxicity to healthy tissue was evaluated using phantoms.

The radiation dose emitted by the linear accelerator was measured by a ionization chamber in conjunction with IONEX 2500/3 equipment. The collimation of the radiation beam was verified by film dosimetry on a polyethylene phantom. The beam

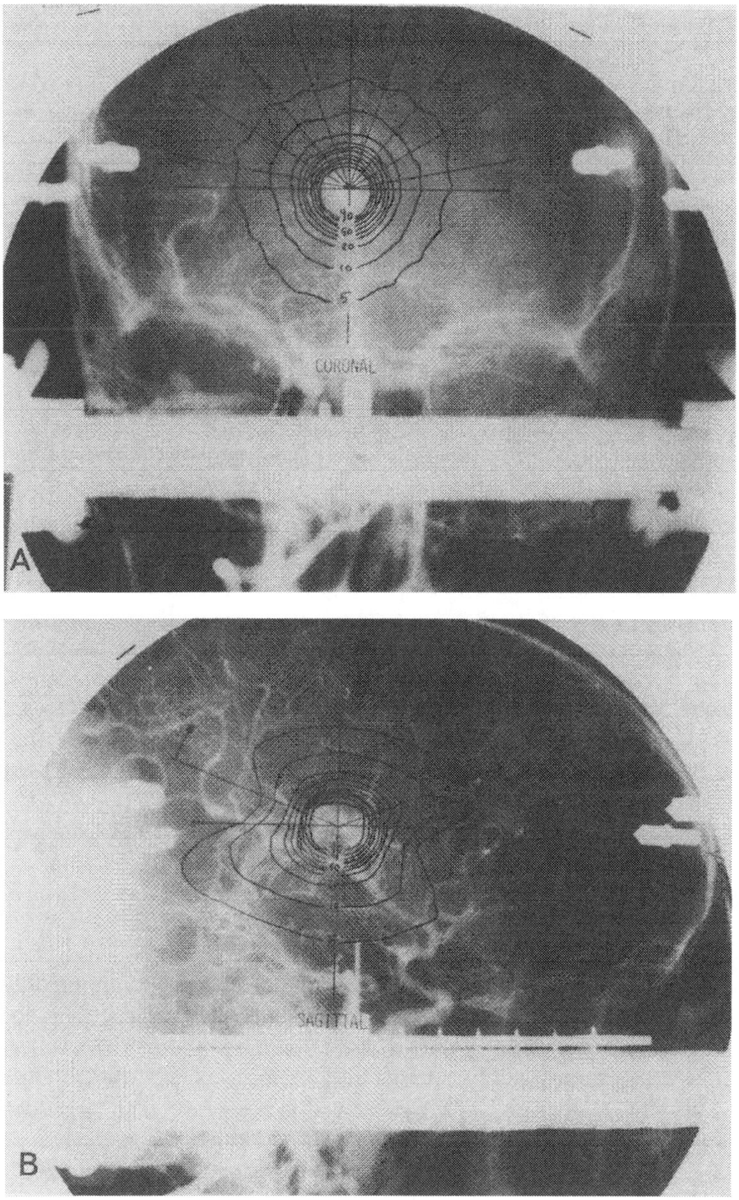

Fig. 1 A, B. Stereotactic angiography of a right-side thalamic arteriovenous malformation with, superimposed, the isodoses obtained with 9 arcs of 140° degrees, field 15 mm × 15 mm. **A** Antero-posterior view: straight lines indicate the planes of arc irradiations. Inner therapeutical isodoses are circular. **B** Lateral view: *arrows* point to the limits of the arc irradiation. Inner isodoses are square with bent curved borders

produced by our Varian Clinac 4 MV linear accelerator has proved to be sharply collimated with virtually no penumbra.

The dose distribution for arc irradiation was calculated by the AECL TP 11 computer system according to J. R. Cunningham's programs and verified by films.

The choice between two available accelerators (4 MV and 18 MV) was made according to dosimetric results obtained with arc irradiations of simulated intracranial targets employing a Alderson Rando anthropomorphic phantom. At the tissue depths usually found in intracranial radiosurgery the 4 MV energy is less toxic to healthy tissues, measured by the ratio

$$\frac{90\%\ \text{isodose area}}{40\%\ \text{isodose area} - 90\%\ \text{isodose area}}$$

The isodoses obtained with multiple isocentric arc irradiations (5–17) were determined by film dosimetry in paraffin phantoms. With the standard built-in square collimators the surfaces of the inner therapeutic isodoses (90%–80%) are slightly barrel-shaped (circular in frontal plane, square with bent curved sides in the axial and sagittal planes). However, for clinical purposes, the isodose shape is not significantly different from a sphere (Fig. 1).

As we have stated in a previous report (Colombo et al. 1985), the steepness of the dose gradient is a function of the amplitude of the spherical sector by which the radiation is administered. By multiplying arc irradiations and angular approaches, dose fall-out is decreased and very steep dose gradients can be achieved.

The standard procedure requires 9–17 arcs of 100°–140°, distributed on a 160° cylindrical sector. With this procedure, normal brain tissue receives less than 5% of the total dose and extracranial sensitive organs less than 0.8‰.

Stereotactic Hardware

We use our original stereotactic apparatus (Colombo et al. 1982, 1984), commonly employed for both functional and nonfunctional stereotactic neurosurgery. This apparatus combines a base ring fixture with an instrument-holder working with a spherical system. Once stereotactic coordinates are set in the remote control, two rotating arms move the instrument-holder on a sphere centered in the target. The target can be reached from any angle by an instrument as long as the radius.

The localization procedures are performed with the base ring fixed to the skull. Even if our procedure could be applied to any type of stereotactic instrumentation, low-profile base ring fixation seems to be more suited to radiosurgery. The base ring must be fixed as low as possible so as to leave the widest possible spherical sector for irradiation. The beam must not strike any metallic parts: to this end we have developed screw holders made of plastic (Fig. 2).

The base ring has the same dimensions as the Riechert-Mundinger stereotactic apparatus. For CT stereotaxy we use the adaptor made by Fisher for our Somatom DR3 CT scan. The localization procedure has been described by Mundinger (Mundinger and Birg 1981; Fig. 3).

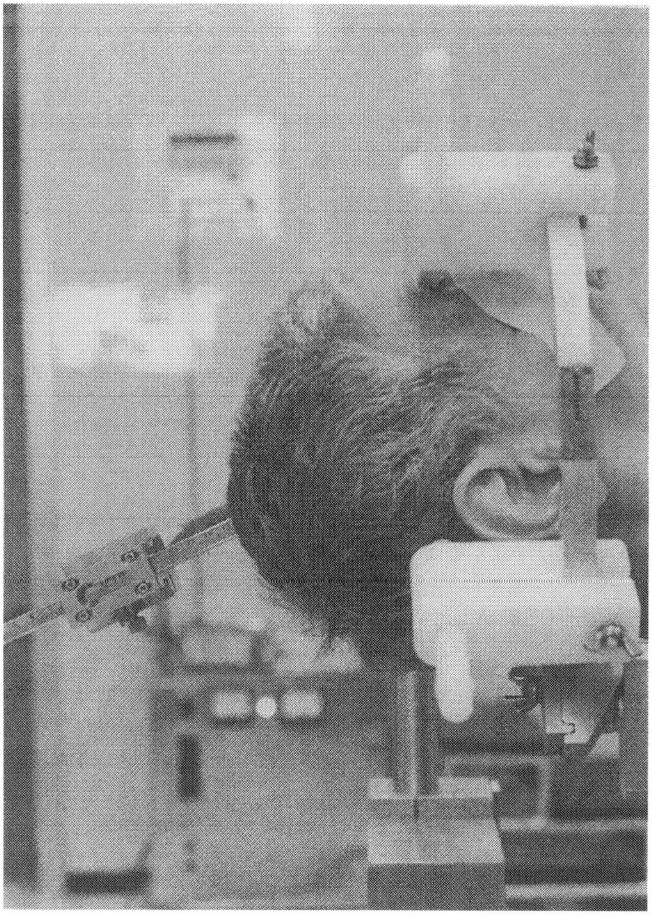

Fig. 2. Patient fixed to the base ring. Plastic fixers and millimetric ruler are clearly identifiable

For targeting arteriovenous malformations, stereotactic angiography is performed (Fig. 1).

Once the localization procedures are completed, a stereotactic biopsy is undertaken if indicated. The aim of the bioptic sampling is not only to ascertain the nature of the lesion but also to verify the shape and size of the target (Ostertag et al. 1980).

After the biopsy a millimetric ruler is fixed to the instrument-holder. The aim of this ruler is to materialize the radiation beam. By moving the ruler, the appropriate planes of irradiation are selected to avoid radiosensitive structures and metallic parts of the fixation frame. With centrally located targets, planes of irradiation are symmetrically disposed at 10–20% intervals on a 160° cylindrical sector. Very low or extracranial targets can be treated by irradiations below the level of the base ring.

The millimetric ruler is also used to record the depths of the tissues to be passed through by the beam before it reaches the target. These data are transmitted to the computer system, which then calculates irradiation parameters.

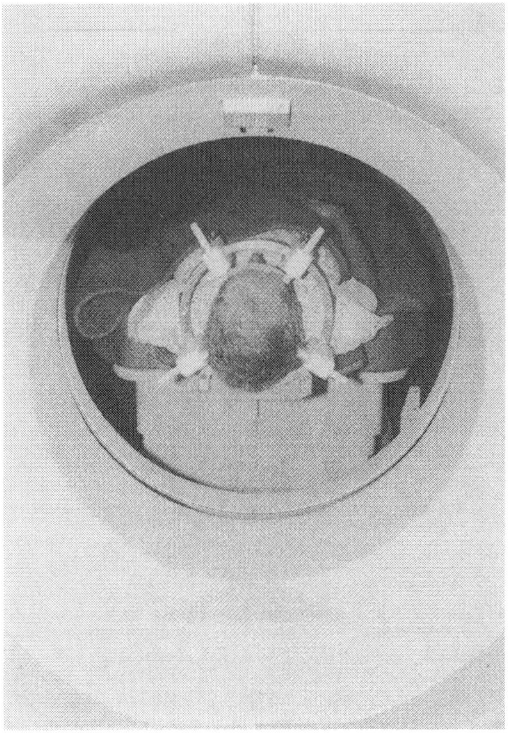

Fig. 3. CT localization procedure

Linear Accelerator Facilities

A headframe adaptor of light alloy has been built for our Varian Clinac 4 MV linear accelerator (Fig. 4). This adaptor stabilizes the base ring or its homologue, the base ring phantom, to the treatment couch.

First the phantom is fixed to the adaptor. The coordinates of the target are set in the phantom's rulers. The treatment couch has 4° freedom. It can be moved along three orthogonal axes; it can also rotate around a vertical axis passing through the isocenter, as most isocentric linear accelerators do. The special feature that makes the machine selected particularly suited for radiosurgical work is the high mechanical accuracy of the movements of the couch (this accuracy has been ascertained in preclinical tests). By controlled movements of the couch, the target is made to coincide with the linear accelerator isocenter, marked by the front pointer. At this moment, brakes are tightened to block the orthogonal movements of the couch. The phantom is replaced by the patient, fixed to the base ring.

Collimator openings are selected for the 80% isodose curve (where the dose gradient is steeper) to coincide with the borders of the target mass. We have a complete series of isodoses obtained with collimator openings from 0.5×0.5 cm to 4×4 cm, but, for reducing toxicity to healthy tissues when dealing with more complicated target shapes, the beam size can be adapted to the variable cross section of the lesion.

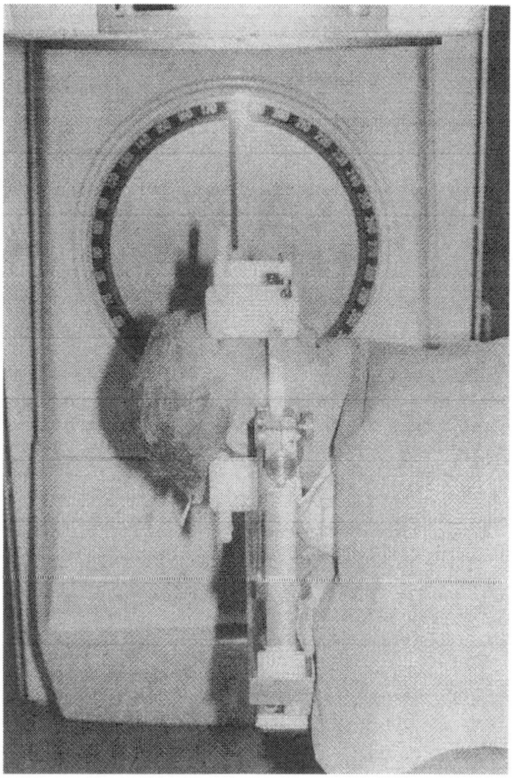

Fig. 4. Linear accelerator radiosurgery. Patient undergoing first arc irradiation in the sagittal plane

Irradiation is usually started on the sagittal plane. After the first arc the couch is rotated by a preselected amount (10–20°) and the arc irradiations are repeated on the preset planes until the total planned dose has been delivered.

Clinical Trial

Fifty-two patients have been treated with single or double shots of between 1000 and 5000 rads. Table 1 shows the patients with follow-up longer than 6 months. Clinical results have been published elsewhere.

Conclusions

Owing to the growing interest in focalized stereotactic irradiation in very recent times, some papers describing techniques for linear accelerator radiosurgery have been published (Betti and Derechinski 1983; Heifetz et al. 1984; Hartmann et al. 1985). In our opinion, the technique we have described has the great advantages of flexibility, simplicity, and efficacy. The solution of multiplying arcs of irradiation by

Table 1. Radiosurgery with linear acceleration: patients treated between November 1982 and December 1984

Glial tumors		22
Low grade astrocytoma	9	
Anaplastic astrocytoma	7	
Glioblastoma	1	
Oligodendroglioma	2	
Ependymoma	2	
Choroid plexus papilloma	1	
Pinealomas		4
Pineocytoma	1	
Pineoblastoma	1	
Germinoma	2	
Medulloblastoma		1
Meningioma		1
Metastasis		2
Craniopharyngioma		2
Pituitary adenoma		1
Arteriovenous malformation		3
Total		36

rotating the patient around a vertical axis passing through the target does away with the need for complicated and expensive machinery such as Betti's swinging chair or the helicoid movement of the radiation source proposed by Heifetz. For the same reason patients can be treated while lying comfortably in a supine position on the treatment couch. The procedure can be extended to patients in poor clinical conditions.

Radiodosimetric tests have demonstrated that with the technique described, spheroidal therapeutic isodoses with very steep dose gradients are obtainable with square fields generated by the linear accelerator's built-in collimator: there is no need of adjunctive collimators, and it is possible to change the field dimensions continuously according to the shape and size of the target mass.

Many centers in the world have in the last decade become deeply involved in the revival of stereotaxy to which the use of computerized imaging techniques has given rise. Our technique may offer to these centers the possibility of extending the principles of stereotaxy to external beam irradiation.

References

Betti OO, Derechinski V (1983) Irradiation stéréotaxique multifasceaux. Neurochirurgie 29:295–298

Colombo F, Zanardo A (1984) Clinical application of an original stereotactic apparatus. Acta Neurochir 33:569–573

Colombo F, Angrilli F, Zanardo A, et al (1982) A universal method to employ CT scanner spatial information in stereotactic surgery. Appl Neurophysiol 45:352–364

Colombo F, Benedetti A, Pozza F, et al (1983) New technique of external stereotactic irradiation. Proceedings of the Congress of the Società Italiana di Neurochirurgia, Bari, 29.9.–1.10.1983

Colombo F, Benedetti A, Pozza F, et al (1985) External stereotactic irradiation by linear accelerator. Neurosurgery 16:154–160

Colombo F, Benedetti A, Pozza F, et al (1985) Stereotactic radiosurgery utilizing a linear accelerator. Appl Neurophysiol

Hartmann GH, Schlegel W, Sturm V, et al (1985) Cerebral radiation surgery using moving field irradiation at a linear accelerator facility. J Rad Onc Bio Phys

Heifetz MD, Wexler H, Thompson R, et al (1984) Single beam radiotherapy knife. J Neurosurg 60: 814–818

Leksell L (1971) Stereotaxy and radiosurgery: An operative system. Thomas, Springfield, Ill

Kjellberg RN, Nguyen NC, Kliman B (1972) Le Bragg peak protonique en neurochirurgie stéréotaxique. Neurochirurgie 18:235–264

Mundinger F, Birg W (1982) CT aided stereotaxy for functional neurosurgery and deep brain implants. In: Driesen W, Brock M, Klinger M (eds) Computerized tomography. Brain metabolism. Spinal injuries. Springer, Berlin, pp 123–134 (Advances in neurosurgery, vol 10)

Ostertag C, Mennel HD, Kiessling M (1980) Stereotactic biopsy of brain tumors. Surg Neurol 14: 257–262

Computer Graphics in Stereotactic Neurosurgery

C. Giorgi, U. Cerchiari, A. Franzini, and G. Broggi

Introduction

Stereotaxy is the surgery of the hidden. The need to visualize the hidden is the reason why simulation by computer graphics technology is necessarily becoming part of this discipline. The combination of these two techniques has been made possible by the development of localizing devices that transfer the position of a target shown by radiological examinations into a surgical frame of reference. Whenever digital diagnostic images are obtained from a patient wearing the stereotactic frame, all data can be expressed with reference to the frame. Suitable localizing devices are now available for the commercial frames: they all generate "artifacts" on the images, from which the orientation of the image plane within the frame coordinates can be derived.

While the value of this achievement is evident for all cases in which a target is demonstrated on diagnostic images [tumors, vascular malformations, and other lesions detected by X-ray or magnetic resonance imaging (MRI)], "functional" targets – those for which stereotaxy was originally introduced – do not directly benefit from this new technology.

The correction of movement disorders or surgery of intractable, chronic pain has to be targeted on diencephalic nuclei which are beyond the resolution of available diagnostic images. Their position is estimated with the aid of stereotactic atlases (Schaltenbrand and Wahren 1977; Andrew and Watkins 1969) and confirmed by intraoperative neurophysiological observations.

Surgery of "Nonvisible" Targets: 3-D Stereotactic Atlas

The use of stereotactic atlases hinges upon the identification of reference points or intracerebral landmarks. These are traditionally provided by positive contrast ventriculography and basically consist of the length and orientation of the intercommissural line. Computed tomography (CT) and MRI can offer a much greater number of intracerebral landmarks, thus reducing the uncertainty caused by individual variability.

Computer graphics are being used to look for morphological analogies between different brains, and so establish the curvilinear coordinate transformation from one brain to another. A solution available at our department (Cerchiari et al. 1986) aims to transfer the borders of anatomical regions or sets of coordinates from one model taken as standard (the atlas) to the individual brains of actual patients.

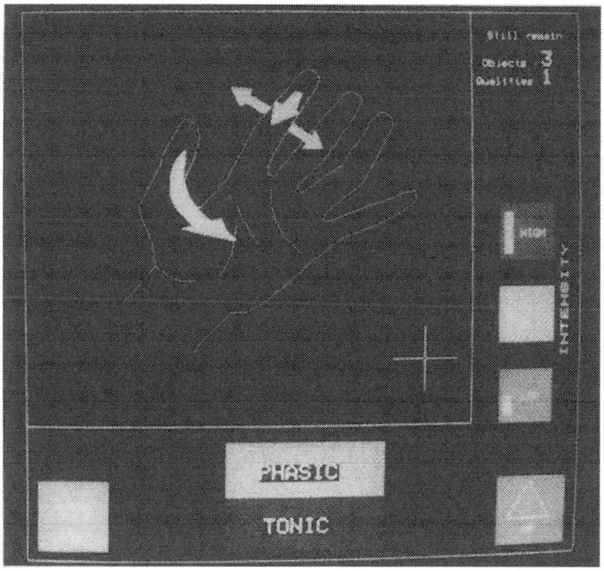

Fig. 1. This image displays symbols corresponding to movements of the fingers, their modalities and intensity. Movements are selected by means of a cursor, and then sent to the computer memory, together with information on the position of the probe and stimulation parameters

Graphics Data Base

Our project is not primarily designed to superimpose anatomical information from the atlas on the individual patient: it aims rather to describe a 3-D transformation that relates single neurophysiological observations, obtained in successive procedures, to a "model brain" in which they acquire statistical significance. The process of matching the two sets of anatomical data is intentionally being confined to the extent to which topology is maintained on different diagnostic images.

Observations gathered during stereotactic functional procedures, consisting of evoked potentials (recorded from the thalamus) and electromyelographic and sensory responses (recorded peripherally after thalamic stimulation), will be scaled to the model anatomy, successive procedures adding to the functional data base, which will eventually constitute a "functional" brain atlas.

A set of graphic subroutines has been devised to symbolize an observation and its location and intensity, in order to store it, together with information on the parameters of stimulation and the position of the tip of the probe. These subroutines use a set of pictures of parts of the body, on which the surgeon can easily move a cursor to select a phenomenon seen in a patient (Fig. 1). The graphic modality speeds up the process of data acquisition and will allow the homogeneous collection of observations.

Clusters of functional observations, stored within the reference of a model brain atlas, are processed efficiently with computer graphics. They are analyzed three-dimensionally and grouped by function, age group, or disease. They can also all be simultaneously analyzed, color- or symbol-coded, and the statistical significance of each can be indicated by color shading.

The surgeon will then establish the target position and the probe trajectory by consulting this functional atlas, which can be supplemented with graphic symbols referring to clinical information such as the frequency of side-effects and clinical follow-up.

The atlas will be adapted to the anatomy of the patient, according to the anatomical information gathered by CT and MRI, obtained stereotactically prior to the procedure. The atlas will then be supplemented with the functional events observed during the procedure by applying the coordinate transformation to the position of the probe at each collection point.

This work is providing momentum for the fascinating work undertaken by other authors (Tasker et al. 1978; Thompson et al. 1977) a few years ago, namely the plotting of neurophysiological responses obtained in humans, in order to describe the distribution of functional pathways in the diencephalon. Computer graphics is much more powerful than the tools then available, quite apart from the fact that the number of observations can now be multiplied through the participation of different neurosurgical centers.

Surgery of "Visible" Targets: 3-D Reconstruction of Cerebral Anatomy

Three-dimensional reconstruction of objects is one of the most attractive features of computer graphics technology. The perception of a solid object can be obtained on a two-dimensional screen by a number of methods, from the rotation of contours with suppression of hidden lines to the more complex technique that adds light and shade to the rotating volumes.

In stereotactic surgery it is not necessary to borrow these computer-aided design technologies; more often than not, the surgeon is satisfied with familiar plain two-dimensional images, provided that the third dimension is known to the computer so that it can calculate the length and trajectory of probes, and operate the necessary rototranslation to deal with different coordinate systems. We have met few cases in which true three-dimensional perception of anatomy is essential, and we feel that the power and simplicity of the stereoscopic approach make it perfectly adequate for the purpose (Szikla et al. 1977).

A good example of its effectiveness is the three-dimensional representation of cerebral vascular anatomy. There have been many approaches to this problem, all attempting to supply the coordinates of each segment of the vascular tree, and consequently burdened with an overwhelming quantity of calculations that inevitably lead to simplification of the vascular tree and to the description of major vessels only (Suetens et al. 1983; Vignaud et al. 1979). The importance of knowing the position of vessels in the closed surgical space of stereotaxy, however, need not be underlined, especially since the vessels themselves can be targets for stereotaxy. Radiosurgery is proving to be a very effective alternative to resection for vascular malformations (Steiner et al. 1974; Betti and Derechinski 1984; Colombo et al. 1985).

This new surgical discipline is going to take advantage of the possibility of precise localization of feeders of extensive vascular malformations. Similarly, the spatial reconstruction of vascular abnormalities induced by the presence of cerebral lesions will facilitate the spread of the very promising stereotactic ablative microsurgery, discussed elsewhere in this book (Kelly et al. 1982, 1986).

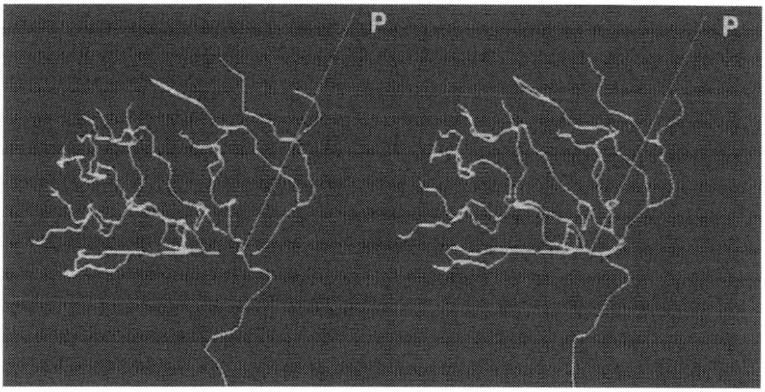

Fig. 2. A pair of simplified angiographic stereoscopic images. The position of a simulated probe *(P)* is indicated. The resulting 3-D image can be "rotated" by displaying successive stereo pairs. The orientation and position of the probe tip can be chosen independently. At every position of the probe the corresponding angles and depth to be entered in the stereotactic frame are given

Our technique utilizes a commercial digital angiography unit, coupled to a small graphic system, and consists of the acquisition of a small series (5–8) of stereoscopic angiographic pairs in stereotactic conditions. The introduction of a localizing device allows orientation of the highly detailed angiographic images in the stereotactic frame of reference. Then, when acquisition is complete and the images are transferred to the graphics controller, the computer can display pairs of digital pictures in a sequence that enables the observer to perceive the three-dimensional image rotating. On those same images the computer can draw stereoscopic pairs of lines, simulating the probe trajectory and its relationship to the vessels. This is shown in Fig. 2, which is a simplified experimental picture that illustrates the graphic output corresponding to a vascular tree, crossed by a stereotactic probe. A chosen trajectory is automtically "translated" into angles and measurements of depth, ready to be entered in the stereotactic system.

Concluding Remarks

The computer graphics world is bristling with hardware and software, and much of both can be adapted to neurosurgical stereotactic procedures. It might very well be that with this marriage, stereotactic neurosurgery will become more widely practised: it is likely that some problems found in traditional neurosurgery may be solved by the stereotactic open technique, with the help of computer graphics technology.

References

Andrew J, Watkins ES (1969) A stereotactic atlas of the human thalamus and adjacent structures. Williams & Watkins, Baltimore

Betti OO, Derechinski VE (1984) Hyperselective encephalic irradiations with linear accelerator. Acta Neurochir [Suppl] (Wien) 33:385–390

Cerchiari U, Del Panno G, Giorgi C, Garibotto G (1986) 3-D correlation technique for anatomical volumes in functional stereotactic neurosurgery. Proceedings of 2nd International Workshop on Time Varying Image Processing and Moving Object Recognition, Florence, September 1986

Colombo F, Benedetti A, Casentini L, Pozza F, Avanzo R, Marchetti C, Chierego G (1985) Stereotactic radiosurgery of intracranial tumors in childhood. J Neurosurg Sci 29:233–237

Kelly PJ, Alker GJ Jr, Zoll JG (1982) A microstereotactic approach to deep-seated arteriovenous malformations. Surg Neurol 17:260–262

Kelly PJ, Bruce AK, Goerss BS, Earnest F IV (1986) Computer assisted stereotaxic laser resection of intra-axial brain neoplasms. J Neurosurg 64:427–439

Schaltenbrand G, Wahren W (1977) Atlas for stereotaxy of the human brain (2nd edn). Thieme, Stuttgart

Steiner L, Leksell L, Forster DM, Greitz T, Backlund EO (1974) Stereotactic radiosurgery in intracranial arteriovenous malformations. Acta Neurochir [Suppl] (Wien) 21:195–210

Suetens P, Jansen P, Haegemans A, Oosterlink A (1983) Three dimensional reconstruction of the blood vessels of the brain from a stereoscopic pair of subtraction angiograms. Image and Vision Computing 1:43–50

Szikla G, Bouvier G, Hori T, Petrov V (eds) (1977) Angiography of the human brain cortex. Springer, Berlin Heidelberg New York

Tasker RR, Hawrylyshyn P, Organ LW (1978) Computerized graphic display of physiological data collected during human stereotactic surgery. Appl Neurophysiol 41:183–187

Thompson CT, Hardy T, Bertrand G (1977) A system for anatomical and functional mapping of the human thalamus. Comput Biomed Res 10:9–24

Vignaud J, Rabischong P, Yver YP, Pardo P, Thural C (1979) Multidirectional reconstruction of angiograms by stereogrammety and computer. Neuroradiology 18:1–7

Superselective Catheterization and Embolization of Cerebral Arteriovenous Malformations

I. S. Choi, A. Berenstein, and E. Flamm

Endovascular embolization of AVMs has progressively been more accepted for the management of lesions that are decided unresectable either because of their size or location. Since the introduction of endovascular treatment for brain AVMs by Leussenhop (Leussenhop and Spence 1960) and for carotid cavernous fistulas by Serbinenko (1974), a variety of catheter assembly systems and embolic materials have widened our possibilities. It is now possible to catheterize very selectively the feeding arteries, which then permits delivery of embolic material into the nidus of the malformation. To decide what treatment is best, in an individual situation, patients are evaluated in a team approach. The neuroradiologists, neurosurgeons, and neurologists, and the patient himself participate in the decision process, let it be surgical resection, embolization, or a combination of both; or in some instances, only conservative management may be best. Certain small size malformations can be successfully treated by radiation therapy (Steiner 1984). This communication will deal exclusively with the embolization part of the treatment.

Goal of Treatment

The goal of treatment, we should emphasize, is prevention of hemorrhage, deterioration of neurological deficit, and if possible restoration of a reversible deficit. Endovascular embolization, either transfemoral or intraoperative (Cromwell and Harris 1980; Debrun et al. 1982) can significantly reduce the size of the nidus of the malformation, and in some instances, even completely obliterate the lesions. In recent years, the combination of embolization, microsurgery or radiation therapy, has improved the possibility of complete obliteration of some of these lesions.

Catheter and Embolic Material

According to the angio-architecture and location of an AVM and the goal of embolization, two major types of embolic materials are available: particulate material such as silicone spheres and Polyvinyl alcohol foam (PVA) of various sizes or liquid agents such as the tissue adhesive, Isobutyl-2-cyanocrylate. The delivery system is may vary and is chosen accordingly (Berenstein and Kricheff 1979a; 1979b).

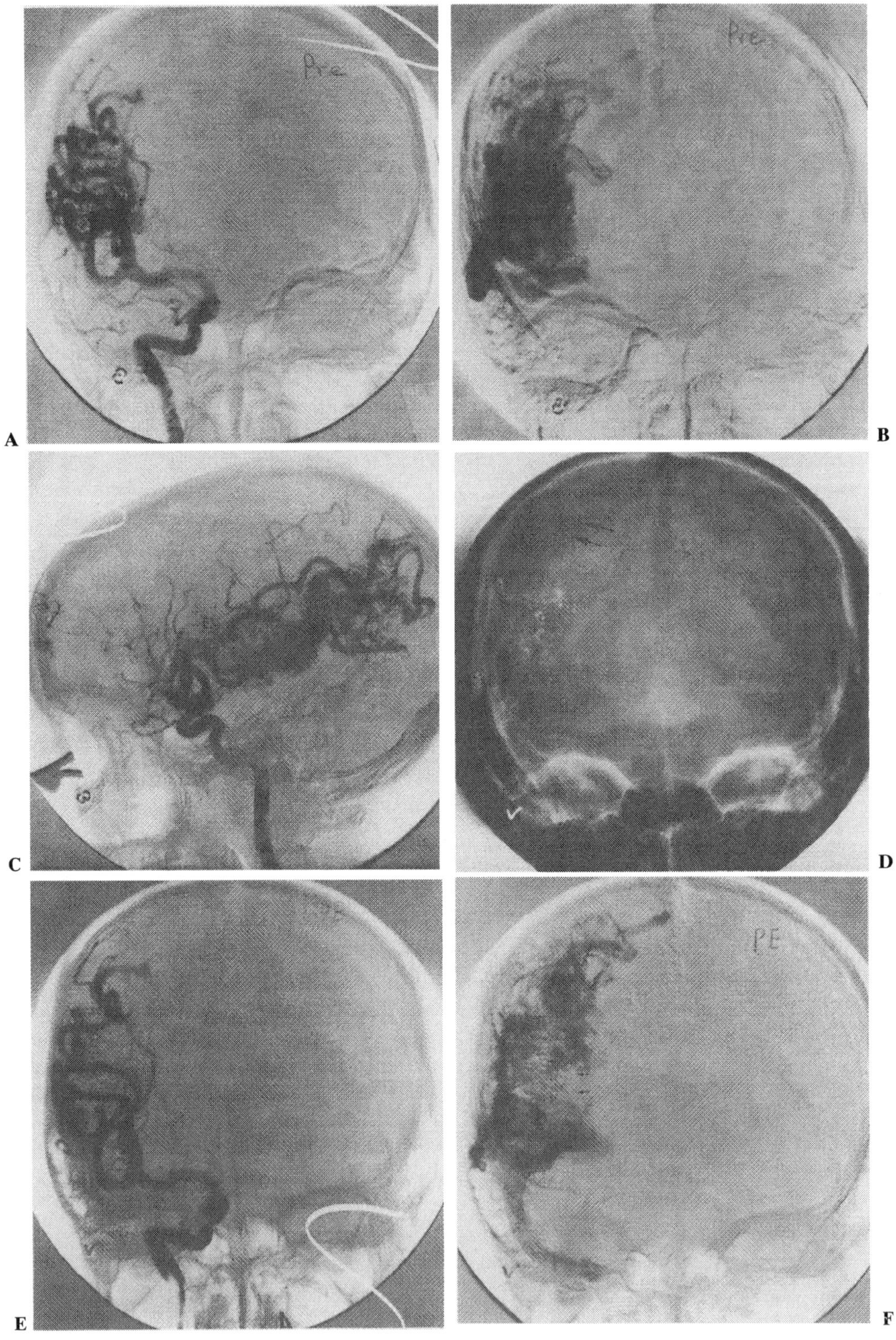
Pre
A
Pre
B
C
D
E
PE
F

Silicone Spheres (Fig. 1)

Silicone spheres are one of the early developed embolic materials which have been used widely since 1960 intraoperatively or transfemorally (Kricheff et al. 1972; Leussenhop and Presper 1975; Leussenhop and Spence 1960). Various sizes are available from 0.5 mm–3 mm. It is most often used as flow directed emboli. The increased blood flow and the larger caliber of the feeding arteries, compared to normal arteries, enables the spheres to be delivered into or close to the nidus of the lesion. The caliber of the feeders and architecture of the nidus as well as the size of fistulae, governs the size of spheres to be used. If preferential flow is not toward the AVM, flow of emboli may be redirected using temporary balloon occlusion of normal arteries (Hilal 1984). Silicone spheres can be effectively used on preoperative embolizations (Stein and Wolpert 1982). Patients with a so called "steal" phenomenon can benefit from this type of embolization. The disadvantage of this technique is that the emboli are injected in the cervical internal carotid artery and there is less control of the final position of the emboli, even if temporary occlusion of normal arteries is performed. Besides, blood may still flow through occluded segments of malformations. Near the end of the embolizations, when hemodynamic changes have occurred, there is a higher chance of occluding normal branches. Therefore, complete obliteration of the nidus is most often not possible. To deliver spheres of different sizes, larger diameter catheter systems are required (Berenstein and Kricheff 1979a; 1979b).

Isobutyl-2-Cyanocrylate (IBCA)

IBCA was first utilized by Zanetti for intravascular occlusion of an aneurysm (Zanetti and Sherman 1972), became useful in cerebral embolization with the introduction of Kerber's calibrated leak balloon catheter system (Fig. 2). The "calibrated" leak balloons are attached to the tip of the very soft silicone tubing with which individual feeding pedicles could be selectively catheterized. Several different catheters systems have become available, with new softer latex balloons introduced by various authors (Debrun et al. 1978; Pevsner 1977). Preferential flow and partial balloon inflation pushes the catheter into the feeders (Fig. 3). Through the leak at the tip of the balloon, radiographic contrast material (Fig. 3A) or liquid embolic agents of low viscosity, such as IBCA (Fig. 3B) could be injected. The quick polymerization property of IBCA in contact with ions or with the change of PH enables the embolic material to produce a cast of the angio-architecture, occluding the nidus, even though there is a fast flow. Inflation of the balloon during injection facilitates polymerization of IBCA in the nidus.

Fig. 1A–F. Example of silicone sphere embolization. **A, B** Frontal views of an early and a late arterial phases of the right internal carotid angiogram; an extensive AVM of the "small vessel" type in the temporo-parietal region fed by the MCA is seen. **C** Lateral view of the right ICA in the same patient. **D** A frontal plain film of the skull shows the distribution of the multiple small silastic spheres. Note, 2 astray spheres *(arrows)*. **E, F** Post embolization frontal views of an early and a late phase. Note, better filling of the normal lenticulostriate arteries (Compared with Fig. 1A) and the reduced nidus of the lesion. Small arrows point to partially subtracted spheres

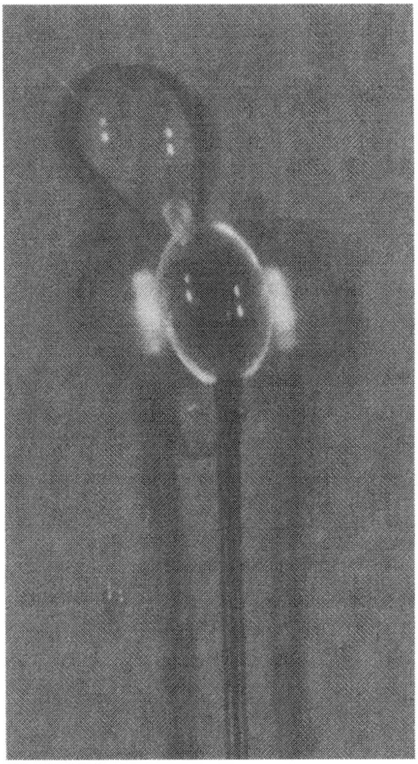

Fig. 2. Calibrated leak microballoon catheter: Kerber's original model. The catheter and balloon are made of silicone rubber. Contrast material or liquid embolic agent can be injected through the leak while inflating the balloon. With the newer latex balloons, one can inject through the leak with or without balloon inflation

Electrophysiologic Monitoring and Provocative Tests

Two broad categories of electrophysiologic monitoring studies (Berenstein et al. 1979a) are employed during interventional procedures. These are evoked potential monitoring, consisting of somatosensory, visual and/or auditory type as well as EEG monitoring during occlusive or provocative studies.

During embolization procedures of intracranial AVMS, amobarbital sodium (amytal) is used to provoke potential neurologic changes prior to embolization of a vascular pedicle. This is a short acting barbiturate. 50 mgms is administered intra-arterially through the catheter prior to embolization. If a neurologic deficit is produced, the effect is immediate and lasts approximately 7 min–10 min. After administration, the patient is monitored for the expected neurologic deficit. Amytal will also produce changes in the evoked potential or in EEG.

Catheterization and Embolization

Selective catheterization of intracranial feeding arteries and its embolization with IBCA are discussed in this section. All transfemoral embolizations are performed

A

B

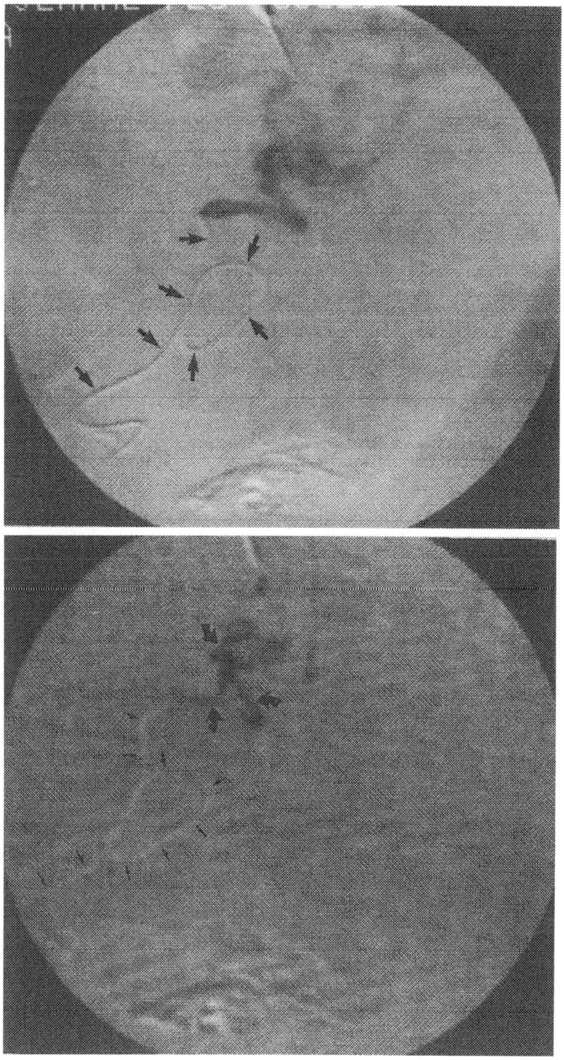

Fig. 3A, B. Superselective catheterization of a parietal branch of the MCA. **A** Lateral digital subtraction angiogram (DSA); the microcatheter can negotiate tortuous curves *(arrows)* and the calibrated "leak" permits injection of the contrast material (B.P. 50 mm Hg mean). **B** DSA of the "in vivo" injection of the radiopaque IBCA *(curved arrows)*. The white curvelinear line represents the already removed microcatheter *(small arrows)*

using introducer sheaths (Berenstein and Kricheff 1979b). It is possible to change the catheter systems without using a guide wire, if the catheter is occluded immediately after delivering embolic materials. 7 French sheaths are commonly used in our institution. Thin wall, 6 French or 6.4 French catheters are first positioned in the cervical portion of the internal carotid or vertebral artery. A calibrated leak balloon catheter is introduced coaxially using pressure chambers (B–D or Ingenor) (Fig. 4). Intermittent inflation of a balloon increases resistance which is pushed into a feeding artery by preferential flow. In some instances, when there is acute angulation of the origin of feeding artery (ACA) (Fig. 5), or small caliber arteries, such as when the choroidal or lenticulostriate arteries are supplying the malformation, two separate

Fig. 4. Assembled delivery system. The B-D pressure chamber is loaded with a calibrated leak balloon catheter *(open arrows)*. By pressure injection of saline through the proximal stopcock *(curved arrow)*, the catheter is propelled coaxially

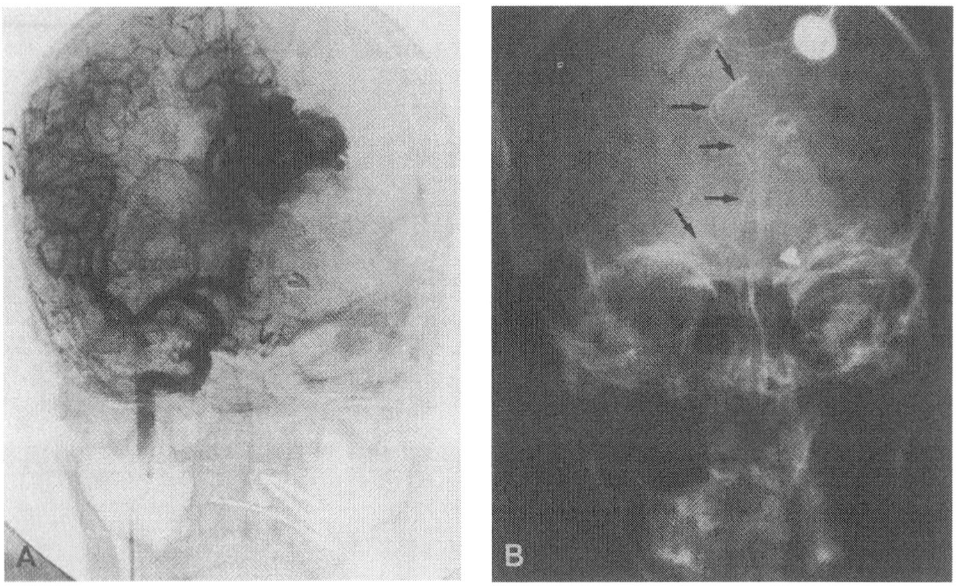

Fig. 5A, B. Frontal subtraction angiogram of the right ICA (**A**). To catheterize the left ACA (**B**), a microballoon was first placed in the right M1 segment, permitting the second microcatheter to enter the ACA *(arrows)*

balloon catheters may be necessary. One balloon is temporarily blocking distal to the origin of the feeder. Therefore, the second balloon catheter can enter into the feeder. The balloon has to be as close to the nidus as possible. The distance from the nidus, angio-architecture of the malformation and presence of normal branches are carefully evaluated by superselective angiogram through the balloon catheter. The superselective angiogram is repeated under systemic hypotension with a mean pressure of 50 mm Hg–60 mm Hg (see Fig. 3A). A larger portion of nidus is often filled

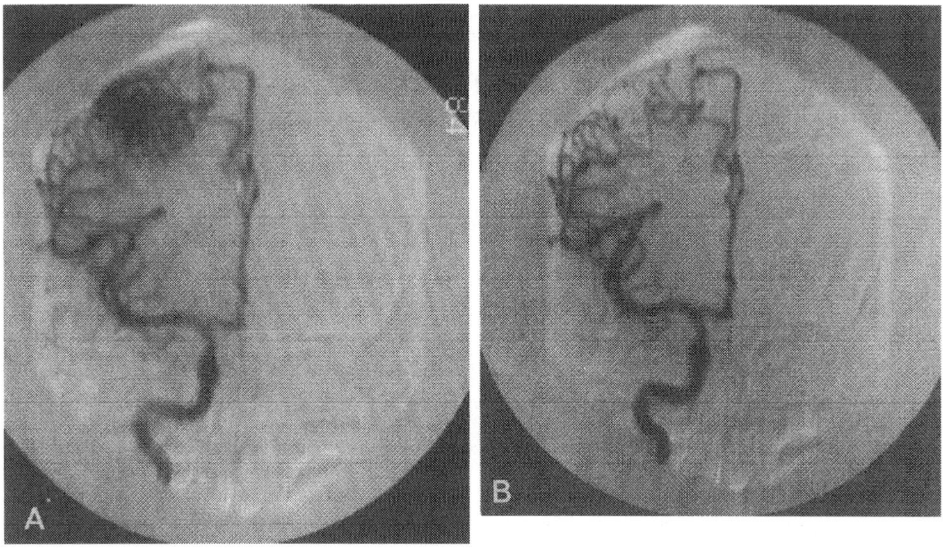

Fig. 6. A Frontal view of the right internal carotid angiogram shows a relatively small AVM in the parietal lobe fed by MCA and ACA, (same patient as Fig. 3). **B** Postembolization control angiogram in frontal view. Complete obliteration of the nidus is demonstrated

through a single pedicle injection under systemic hypotension in monocompartmental lesions that may have multiple feeders (Fig. 6). Besides, slower blood flow gives better control of injection of IBCA mixture, and less chance of venous occlusion or pulmonary embolism.

The polymerization time of IBCA can be delayed by adding pantopaque up to 7s–8s (Cromwell and Harris 1980). Recently, Vinuela (1985) studied accurate determination of polymerization time by adding acetic acid. We use tantalum powder which is an inert metal to opacify the IBCA mixture and monitor the actual injection under fluoroscopy (see Fig. 3B). The use of digital subtraction angiography has improved remarkably our capability of more accurate catheterization in a shorter time and more precise injection of the embolic material.

Staged Embolizations

The majority of arteriovenous malformations referred to for endovascular treatment are large lesions with multiple feeders. Abrupt changes of cerebral perfusion may occur when occluding multiple feeders in one sitting, which may produce a break in normal perfusion pressure (Spetzer et al. 1978) and subsequent cerebral edema and hemorrhage. For large lesions, 2 or 3 feeding arteries are embolized in one sitting, which may be repeated after 4–5 days. By sufficient reduction of the nidus after embolization, AVMs may be converted to surgically resecteble lesions, or may be small enough for radiation therapy. Some patients who presented only with seizure disorder or neurological deficit may then be managed conservatively.

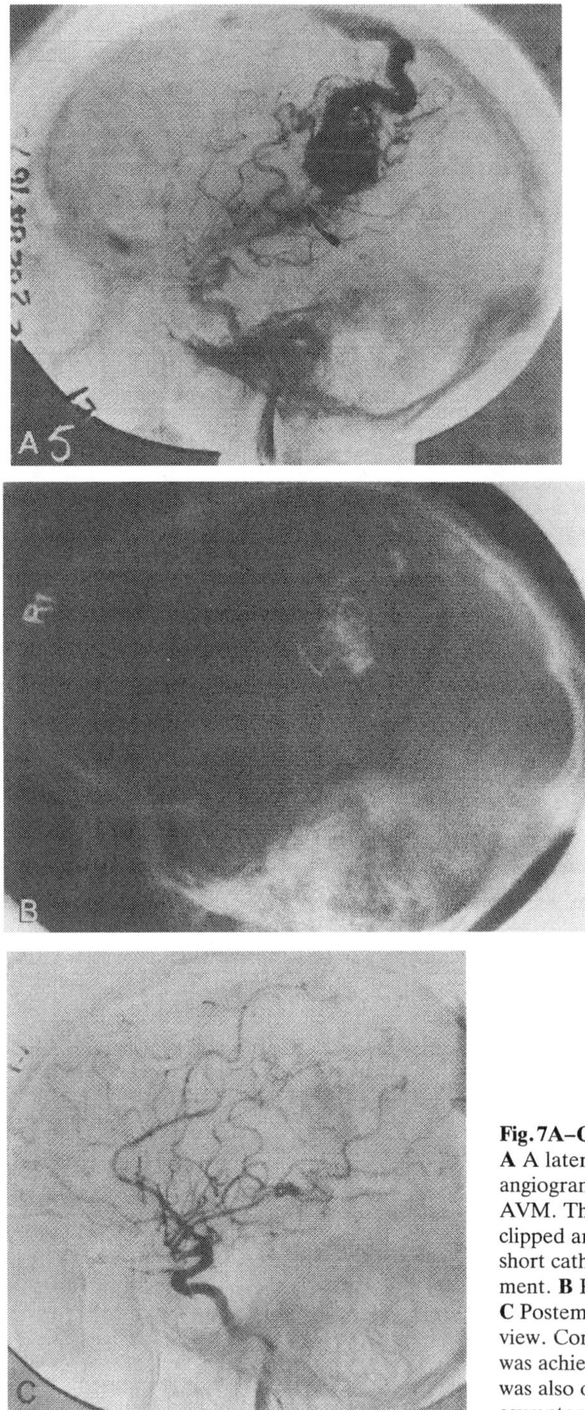

Fig. 7A–C. Intra-operative embolization. **A** A lateral view of the left internal carotid angiogram. Note, two MCA branches of the AVM. The main feeding artery *(arrow)* was clipped and IBCA was injected through a short catheter, inserted in the distal segment. **B** Radiopaque IBCA cast obtained. **C** Postembolization angiogram in lateral view. Complete obliteration of the nidus was achieved. The superior sagittal sinus was also occluded, but the patient was asymptomatic

Combined Treatment

Team effort is essential for management of cerebral arteriovenous malformations. In spite of improvement of surgical and endovascular techniques, it is still difficult to irradicate large lesions by a single mode of treatment. Pre-operative embolization increases the possibility of total surgical resection of large AVMs. By embolizing deep seated feeding pedicles of the anterior or posterior cerebral arteries, it can convert an unresectable AVM to a resectable one. Sometimes, embolization can completely obliterate post-surgical residual portion of the nidus which could not be removed either because of critical locations or the inability to reach the nidus. Not all feeding arteries are reachable by endovascular routes. The small caliber, difficult curves or relatively slow flow prevents migration of a balloon catheter toward the nidus. Normal arteries, which may supply a critical region may originate from the same pedicle as an AVM feeder. On these occasions, embolization is not possible to obliterate the whole nidus. The remaining nidus can then be treated by radiation therapy or even surgery if possible.

The anterior cerebral artery or the middle cerebral artery supply which could not be reached by transfemoral route can be exposed surgically and IBCA can be injected near the nidus (Fig. 7) (Cromwell and Harris 1980).

Results and Complications

The development of better catheter systems and the advancement of catheterization techniques enabled us to improve the results of embolization and reduce the number of complications. We have treated 126 patients excluding dural arteriovenous malformations from 1978 to 1984. A total of 193 procedures were performed. We are in the process of a detailed analysis of our results. Preliminary statistics of our first 98 patients (1983), includes: 42 patients treated with silicone spheres alone or in conjunction with IBCA. Most of these patients were treated during our early experience. 56 patients were treated only with IBCA. Follow up from 2 to 6 years has shown that 3 patients with silicone sphere embolizations have rebled, while only 1 patient treated with IBCA embolization has had a repeat episode of bleeding. 5 patients had complete obliteration of the AVM with IBCA. There was a complication rate of 15.3% including all transient neurological deficits. 5 deaths directly caused by or related to embolization had occurred. 3 patients died probably of hemorrhage. One patient died of "perfusion pressure breakthrough" caused by complete occlusion of the malformation in one sitting. The 5th patient died of retrograde thrombosis of transverse pontine arteries after embolization of a medial left temporal AVM. 63 patients were treated palliatively because of their sizes or locations. Clinical improvement was observed in the majority of our patients.

References

Berenstein A, Kricheff II (1979a) Catheter and material selection for transarterial embolizations: Technical considerations. Part I: Catheters. Radiology 132:619

Berenstein A, Kricheff II (1979b) Catheter and material selection for transarterial embolizations: Technical considerations. Part II: Materials. Radiology 132:631
Berenstein A et al (1984) Somatosensory evoked potentials during spinal angiography and therapeutic transvascular embolization. J Neurosurg 60:777
Cromwell LD, Harris AB (1980) Treatment of cerebral arteriovenous malformations. A combined neurosurgical and neuroradiological approach. J Neurosurg 52:705
Debrun G et al (1978) Detachable balloon and calibrated leak balloon techniques in the treatment of cerebral vascular lesions. J Neurosurg 49:635
Debrun G et al (1982) Embolization of cerebral arteriovenous malformations with bucrylate. Experience in 46 cases. J Neurosurg 56:615
Hilal SK (1984) Endovascular treatment of AVMS of CNS. In: Wilson CB, Stern BN (eds) Intracranial arteriovenous malformations. Williams and Wilkins, Baltimore London, pp 259–273
Kerber C (1976) Balloon catheter with a calibrated leak: A new system for supserselective angiography and occlusive catheter therapy. Radiology 120:547
Kricheff II et al (1972) Transfemoral catheter embolization of cerebral and posterior fossa arteriovenous malformations. Radiology 103:107
Leussenhop AJ, Spence WT (1960) Artificial embolization of cerebral arteries. Report of use in a case of arteriovenous malformation. JAMA 172:1153
Leussenhop AJ, Presper JH (1975) Surgical embolization of cerebral arteriovenous malformations through internal carotid and vertebral arteries. Long term results. J Neurosurg 42:445
Pevsner PH (1977) Microballoon catheter for superselective angiography and therapeutic occlusion. AJR 128:225
Serbinenko FA (1974) Balloon catheterization and occlusion of major cerebral vessels. J Neurosurg 41:125
Spetzer RF et al (1978) Normal perfusion pressure breakthrough theory. Clin Neurosurg 25:651
Stein BM, Wolpert SM (1982) Arteriovenous malformations of the Brain II: Current concept and treatment. Arch Neurol 37:69
Steiner L (1984) Treatment of arteriovenous malformations by radiosurgery. In: Wilson CB, Stein BM (eds) Intracranial arteriovenous malformations. Williams and Wilkins, Baltimore London, pp 295–313
Vinuela F (1985) Accurate determination of polymerization of IBCA by adding acetic acid: Presented at the NYU Postgraduate Course in Interventional Neuroradiology
Zanetti PH, Sherman FE (1972) Experimental evaluation of a tissue adhesive as an agent for the treatment of aneurysms and arteriovenous anomalies. J Neurosurg 36:72

Comparison of Magnetic Resonance Imaging with Computed Tomography in the Detection and Characterization of Intracranial Tumors

T. H. Newton, W. M. Kelly, M. Brant-Zawadzki, and D. Norman

The rapid development of clinical magnetic resonance imaging (MRI) and its recognition as a valuable tool in neurodiagnosis have generated excitement and enthusiasm rivalling that associated with the introduction of computed tomography (CT) a decade ago. Magnetic resonance imaging has several advantages over X-ray CT. The multiple physical properties of tissue that determine image contrast in MRI permit superior detection and characterization of altered tissue. The relatively small alterations in focal water concentration caused by many brain disorders produce proportionally larger alteration in intensity of the MRI signal than of the CT signal. Differential sensitivity of MRI to resonating protons found in other molecular substrates, such as fat or hematoma, may yield additional information of histologic significance. The ability of MRI to visualize blood vessels provides anatomic landmarks in normal patients and assumes diagnostic importance in the presence of vascular lesions. Finally, MRI has no known biologic hazards, iodinated intravenous contrast agents are not required, bone artifacts are eliminated, and images may be obtained directly in any plane.

Early reports have already indicated the greater sensitivity of MRI for the detection of a wide variety of intracranial abnormalities including neoplastic, congenital, inflammatory, demyelinating and vascular diseases (Holland et al. 1980; Young et al. 1981; Crooks et al. 1982a; 1982b; Buonanno et al. 1982; Bydder et al. 1982; Brant-Zawadzki et al. 1983).

Many disorders of the brain, including tumors, are accompanied by a focal increase in water concentration, either as a primary or secondary phenomenon. These changes tend to produce 1) prolongation of T1 relaxation time, and 2) a prolongation of T2 relaxation, and 3) an increase in the resonant proton density relative to normal brain tissue. Because the latter two characteristics are directly proportional to signal intensity, while the former is inversely proportional, their combined effects on overall signal intensity may cancel each other out, thus abolishing contrast between disparate tissues. The proper choice of instrument parameters is therefore critical in detecting differences in signal intensity (i.e., contrast) between normal and pathologic tissue.

The T2-weighted image is most likely to provide optimal contrast sensitivity for lesions with increased water content by highlighting resonant proton density and T2 relaxation effects while diminishing the influence of T1 relaxation. The effect of differential T1 relaxation rate of protons in different tissue can be reduced in displayed images on spin echo MRI studies obtained with relatively long TR intervals. When TR intervals exceed 3 to 4 times the T1 time constant for most tissues in a volume being imaged, ample time elapses for the magnetization of "slower" tissues (long T1)

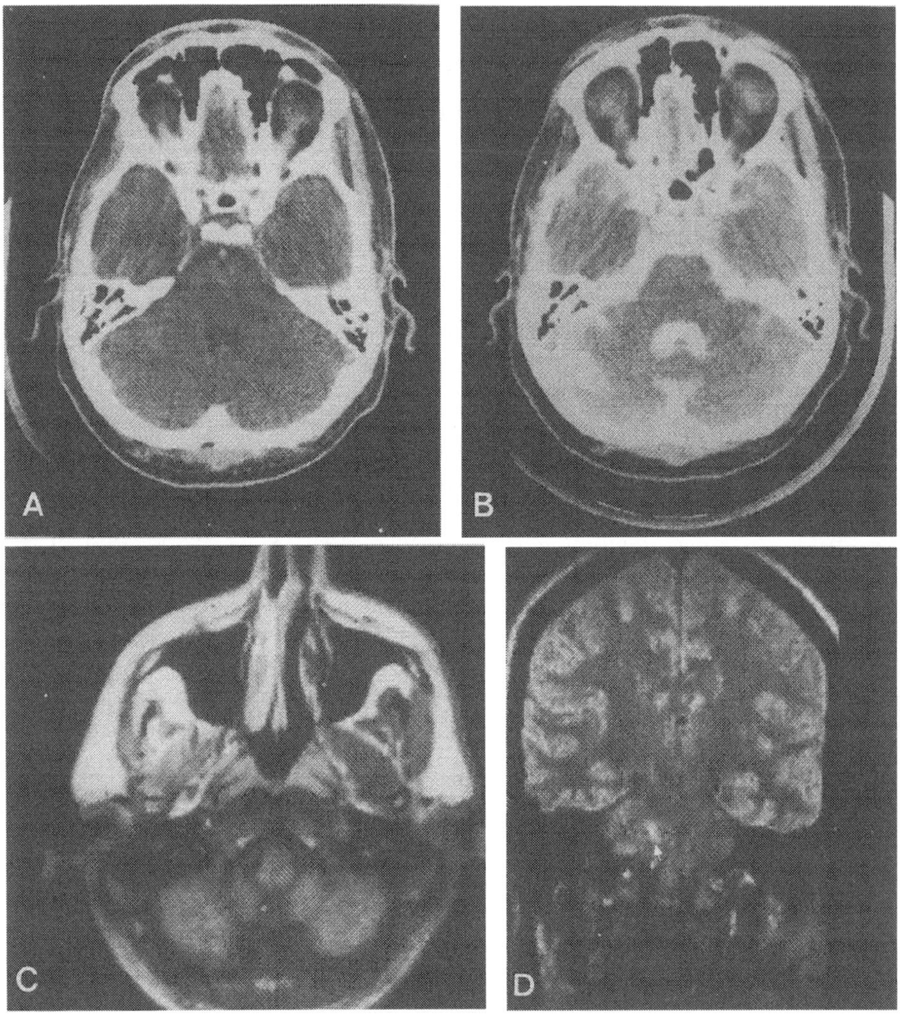

Fig. 1A–D. Astrocytoma of brainstem in 23-year-old with right sided VII cranial nerve palsy and positive evoked responses. **A** Contrast-enhanced CT scan showed no abnormality of the brainstem. **B** Metrizamide CT cisternogram showed normal contour features at the level of the pons. **C** Axial MR image demonstrated denervation atrophy of muscles in the right infratemporal fossa, with fatty replacement giving rise to increased signal. **D** Coronal MRI (T2-weighted) image revealed the lesion *(arrow)* on the right side of the pons

Fig. 2A–E. "Aqueductal stenosis" due to midbrain astrocytoma in patient with neurofibromatosis. **A, B** Computed tomographic sections through midbrain show only ventricular dilation without any attenuation abnormality. **C** Magnetic resonance image at level of dilated ventricles better demonstrates interstitial edema surrounding the ependyma as a result of obstructive hydrocephalus. **D** The midbrain lesion *(arrows)* is clearly visible as a bright focus just behind the aqueduct on a sagittal T2-weighted MR image (due to relative prolongation of T2). **E** T1-weighted image shows lesions as a dark area, consistent with a long T1

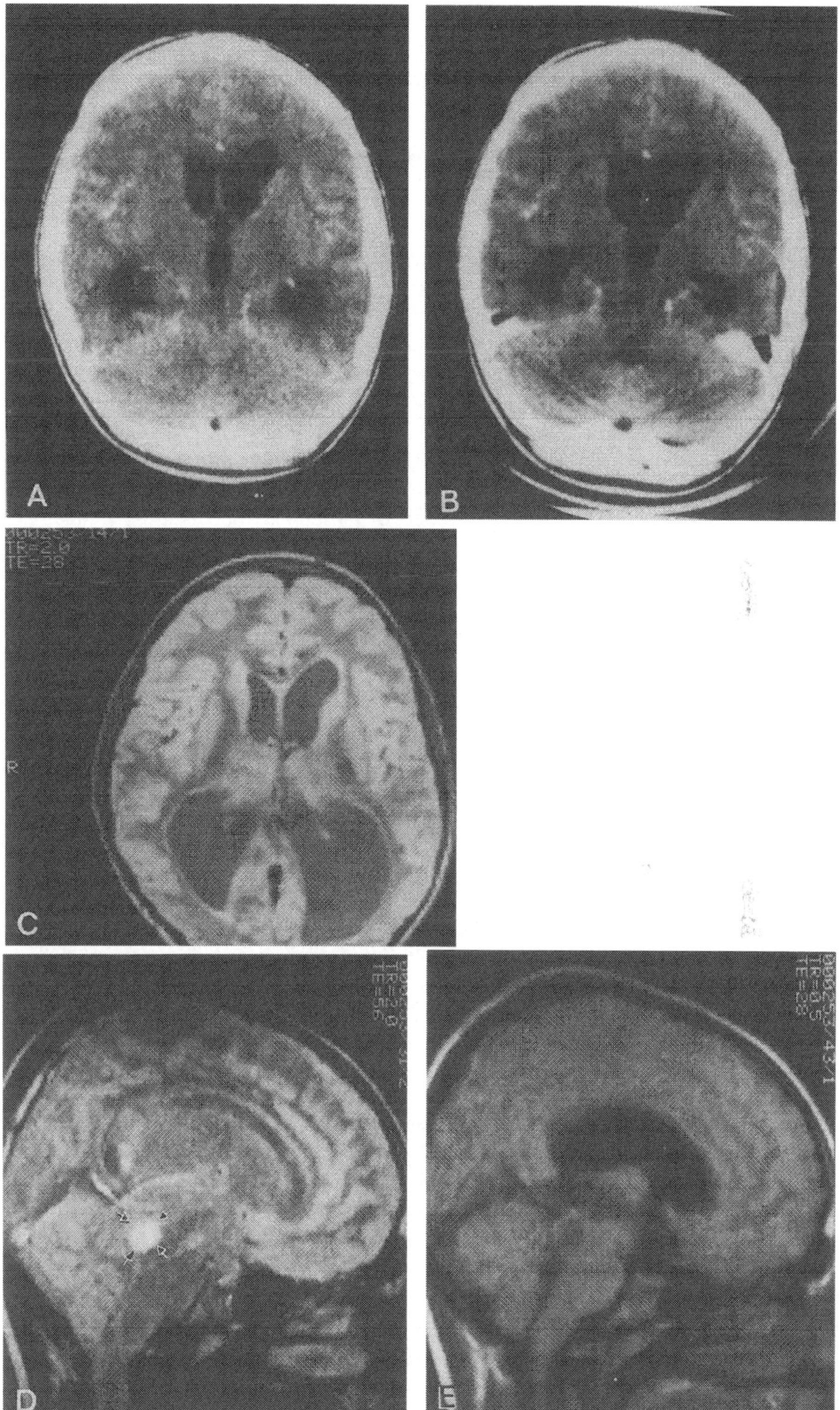
A
B
TR=2.0
TE=28
R
C
D
E

to "catch up" with the more rapidly magnetizing tissues (short T1). Thus, at the moment of spin echo sampling, although the component of the signal amplitude based on T1 dependent relaxation phenomenon is nearly maximal, it is also very nearly uniform in all the different tissues in the image (exceptions include cerebrospinal fluid, vitreous, and fluid collections which have very long T1 relaxation times). With such a choice of long excitation interval (TR), T1 differences are minimized and contribute little, if any, to image contrast. Under these circumstances, the components of the signal amplitude based on resonating proton density and T2-dependent relaxation phenomenon are "unmasked" and are the major determinants of differential signal intensity in distinct tissues, i.e., MR image contrast.

Once the presence of a lesion is detected with a T2-weighted spin echo sequence, the abnormality can be further characterized with a T1-weighted (short TR) sequence. By choosing a short TR interval of 500 ms, for instance, T1 relaxation occurs only partially in normal or abnormal tissues where most T1 time constants are in the range of 400 ms–1000 ms (at a field strenght of 0.35 T). The differential magnetization (T1 relaxation) of distinct tissue types results in heterogeneous amplitudes of signal intensity and provides contrast between them in the resulting image. For example, fat or hemorrhage, which typically has a relatively rapid T1 relaxation, will show relatively bright signal strength. Such a short TR, i.e., "T1-weighted" sequence may also permit discrimination of tumor nidus tissue from the surrounding secondary edema. Distinction between tumor margin and adjacent brain edema on MRI depends on the detection of a relatively prolonged T1 relaxation of a tumor nidus compared with adjacent edematous and/or normal tissues. The tumor nidus would, therefore, appear as a "darker" region on T1-weighted spin echo images.

In a recent review of 84 patients being evaluated for intracranial neoplastic disease, MRI was more sensitive than CT in 21 patients in whom CT of the brain was either normal (Fig. 1), showed only ventricular enlargement (Fig. 2), or demonstrated diffuse, often subtle mass effect without abnormal enhancement (Brant-Zawadzki et al. 1984). Several of the nondiagnostic CT examinations were compromised by suboptimal scan quality due to streak artifacts. In 54 (84%) of the remaining 63 patients, MRI more clearly indicated the extent of abnormal tissue.

In all 79 of the positive MRI studies, altered tissue was best detected by using the long TR (T2-weighted) spin-echo screening technique. The contrast was generally greater on the second echo (56 ms) image. Deep hemispheric mass effect, such as buckling of white matter tracts, was easily recognized by the excellent differentiation of gray and white matter, even when accompanied by little or no signal abnormality.

The short TR (T1-weighted) spin echo images were not as sensitive and failed to depict the abnormality in 32% of the patients. In the remaining 68%, the T1-weighted images provided information which permitted further characterization of lesions. This additional information included locating the epicenter of the tumor nidus (Fig. 3)

Fig. 3A–F. Diffuse astrocytoma in patient with temporal lobe symptoms. **A, B** Computed tomographic scans revealed both a temporal lobe lesion (note effacement of temporal horn in *A*) and an occipital lobe abnormality (anterior displacement of trigonal choroid in *B*) on the left, suggesting metastatic disease as the most likely diagnosis. **C–E** Magnetic resonance images showed contiguity of the left hemispheric abnormality through the deep white matter (**D**). **F** Biopsy of the occipital tumor focus indicated by dark area on this T1-weighted image confirmed astrocytoma. "Astrocytic change" was present on the contiguous white matter

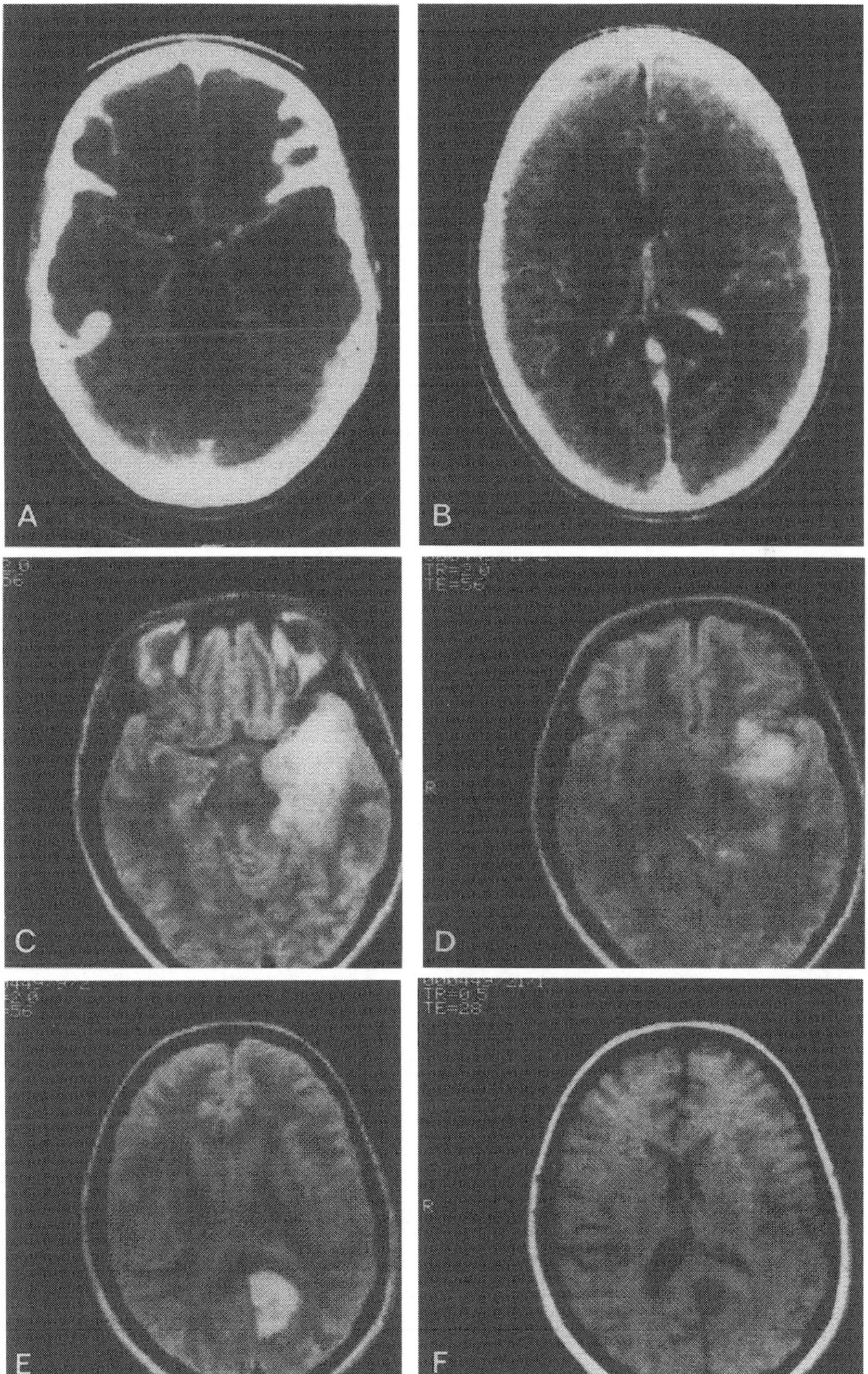
A
B
TR=3.0
TE=56
C
D
TR=0.5
TE=28
E
F

and detecting of fatty tissue component or subacute and old tumoral hemorrhage. Magnetic resonance imaging was far more sensitive than CT in the detection of hemorrhagic foci.

With the exception of meningiomas, peripheral lesions were better visualized and more confidently identified as being intra- or extra-axial on the basis of MRI findings, particularly when located near or adjacent to the skull base. In such locations, MRI provided sharp contrast between normal cerebrospinal fluid spaces and adjacent structures and depicted vascular channels.

Small foci of punctate or curvilinear calcifications noted on CT were not detected by MRI in 7 patients. The paucity of MRI signal arising from normal cortical bone also limits the visibility of osseous landmarks and may initially be troublesome when planning surgical approaches near the skull base. For similar reasons, subtle foci of bone erosion may be missed with MRI.

5 lesions were not detected on MRI. These included 3 pituitary microadenomas and one intracanalicular neurinoma (revealed by high-resolution thin-section 1.5-mm CT). In addition, a focus of ependymal tumor spread seen with CT was masked by MRI signal abnormalities from radiation damage (the latter was missed with CT).

Summary

The ultimate role of MRI in diagnosis of brain tumor remains to be defined. Its clinical application will depend on its ability to detect and characterize neoplastic disease and to evaluate the effects of therapy efficiently and economically compared with CT. Combined T2- and T1-weighted sequences appear to secure maximal sensitivity and specificity in the detection and characterization of intracranial neoplastic disease. Magnetic resonance imaging will, therefore, probably replace CT as the primary diagnostic imaging modality for initial and follow-up evaluations of patients with intracranial neoplastic disease.

References

Brant-Zawadzki M, Davis PL, Crooks LE, et al (1983) NMR demonstration of cerebral abnormalities: comparison with CT. AJNR 4: 117–124

Brant-Zawadzki M, Badami JP, Mills CM, Norman D, Newton TH (1984) Primary intracranial tumor imaging: a comparison of magnetic resonance and CT. Radiology 150: 435–440

Buonanno FS, Pykett JL, Brady TJ, et al (1982) Clinical relevance of two different nuclear magnetic resonance (NMR) approaches to imaging of a low grade astrocytoma. J Comput Assist Tomogr 6: 529–535

Bydder GM, Steiner RE, Young IR, et al (1982) Clinical NMR imaging of the brain: 140 cases. AJNR 3: 459–480

Crooks L, Arakawa M, Hoenninger J, et al (1982a) Nuclear magnetic resonance whole-body imager operating at 3.5 kGaus. Radiology 143: 169–174

Crooks LE, Mills CM, Davis PL, et al (1982b) Visualization of cerebral and vascular abnormalities by NMR imaging: the effects of imaging parameters on contrast. Radiology 144: 843–852

Holland GN, Hawkes RC, Moore WS (1980) Nuclear magnetic resonance (NMR) tomography of the brain: coronal and sagittal sections: J Comput Assist Tomogr 4: 429–433

Young IR, Burl M, Clarke GJ, et al (1981) Magnetic resonance properties of hydrogen: imaging the posterior fossa. AJR 137: 895–901

Advances in Neurostimulation Devices

J. Siegfried and G. Rea

The past decade has seen increased use of devices designed to stimulate along the spinal axis or within the brain substance to treat intractable pain or disabling movement disorders. Concomittent with this period of expanding use, refinements have been made both in the electrodes and the stimulation devices. Some of the more significant alterations will be discussed.

Electrodes

Regardless of whether an electrode is destined for the epidural space of the spinal canal or the deep gray matter of the brain, there are some universally desired characteristics. These include (1) the lack of toxicity after long term stimulation, (2) positional stability after placement, and (3) easy insertion and removal (Brummer and McHardy 1977). In attempting to achieve these general characteristics, some specific alterations according to placement site have also been made.

Spinal Axis Electrodes

The possible toxic effects of different metals on neural and surrounding tissue, with and without electric current, have been studied. The metals such as platinum, iridium, rhabdium, stainless steel, and tantalum were found to have essentially no toxicity (Brummer et al. 1983; Donaldson et al. 1985; Dymond et al. 1970; Mortimer et al. 1970; Pudenz 1977). Accordingly, many electrodes are now of platinum or platinum-iridium alloys which show very little metal dissolution and a potential lifetime of decades.

Positional stability of the electrodes was not a problem with the early attempts at spinal cord stimulation, since a laminectomy was required and the electrodes were sutured into place. Although subsequent development of a percutaneous electrode placement method gave greater ability to intra- and post-operatively test the patient's response to stimulation, it does not fix the electrodes into place. Therefore, electrode migration and the resultant change in therapeutic response can become a significant problem. Avery and colleagues (1980) attempted to decrease the possibility of migration by placing Dacron flocking on the electrode near the stimulating sites (Fig. 1). They found that this effectively stopped the migration. Although it also required approximately a 10 times greater force to remove the electrode, this could

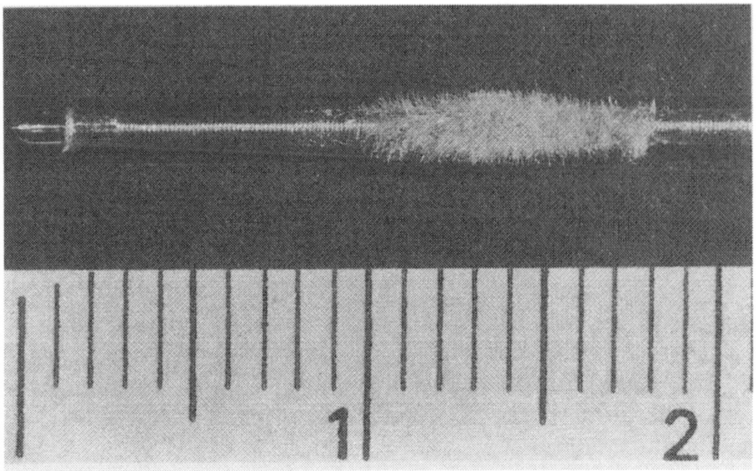

Fig. 1. Avery electrode with addition of Dacron flocking to a segment near the stimulating tip in order to promote tissue ingrowth to improve chronic stability (measurement in cm's)

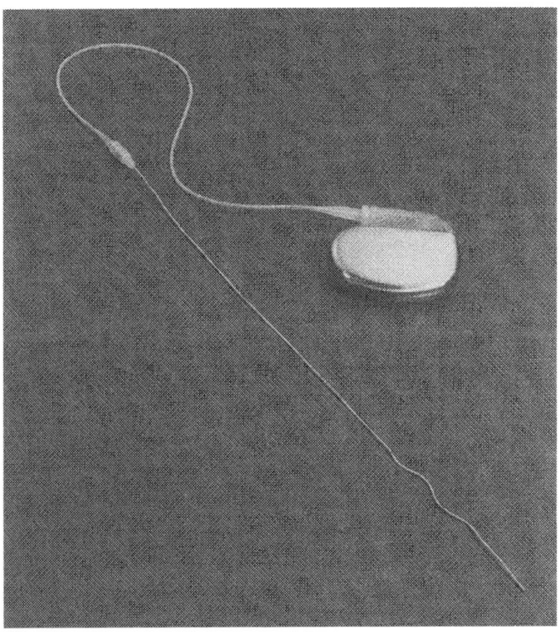

Fig. 2. Total implant (programmable pacemaker) coupled with an epidural sigma-electrode (Medtronic)

still be done without damage to the dura. Medtronic developed an electrode now widely used which shows some ondulation after removal of the stylet (due to this aspect, this electrode is called a "sigma electrode") (Fig. 2 and 4). Blume et al. (1982) also reported on their use of Dacron flocking, but because they felt it decreased the ease of withdrawing the electrode while the needle remained in place, its

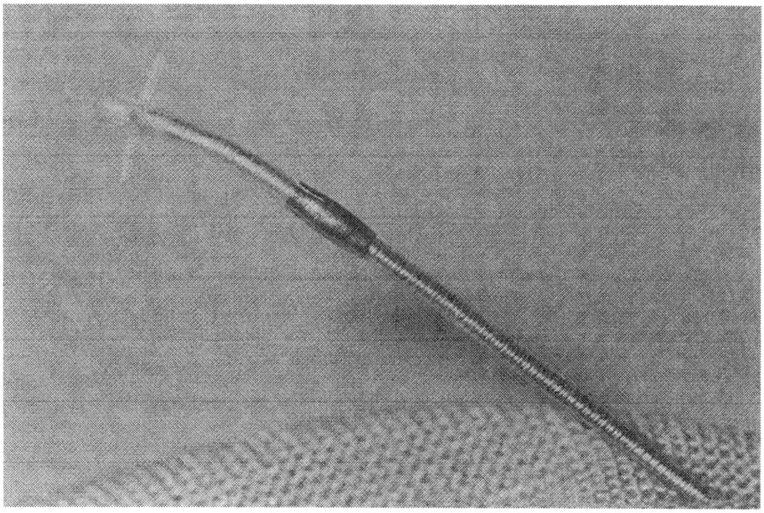

Fig. 3. Prototype of an electrode designed by Martinez with 4 Silastic spurs which radiate at 90° angle for anchorage in the epidural space (by courtesy of Dr. Martinez)

use was abandoned. Another electrode alteration designed to fix it in the epidural space is currently being evaluated by Martinez (1985). He is using a specially designed lead with 4 silastic spurs which radiate at 90° angles from the electrode's tip (Fig. 3). The use of multiple contact electrodes may also be helpful in overcoming the migration problem, since a different stimulating point might be effective. These multiple contact electrodes, like the Quad Medtronic with 4 contacts, can be introduced through a Touhy needle like the monopolar ones or through a small laminectomy. Each of the 4 contacts can be independently selected via an external radiofrequency stimulator (Fig. 4). However, at the current time there is no ideal solution to the problem of spinal epidural electrode migration, and further study and evaluation in this area is required.

Probably the most significant advancements in electrode design have dealt with their size and subsequent ease of placement. As electrodes became smaller, a laminectomy was no longer required, percutaneous placement could be done, and even multiple contact points were possible. These improvements, coupled with evidence that monopolar stimulation is as effective as bipolar (Cobun 1980; Comte and Siegfried 1979; Comte 1982), have encouraged and allowed functional neurosurgeons to place them under local anaesthesia. This not only significantly decreases the risk of the procedure, but also increases the chance of therapeutic success which is more likely with stimulation-induced paresthesias in the painful areas (Law and Muller 1982; Sharkey et al. 1982).

Currently at the University of Zurich, percutaneous placement of monopolar electrodes in the spinal epidural space has been performed in 520 consecutive patients. The electrodes with Dacron flocking are used in exceptional patients who have several episodes of electrode migration resulting in loss of therapeutic stimulation.

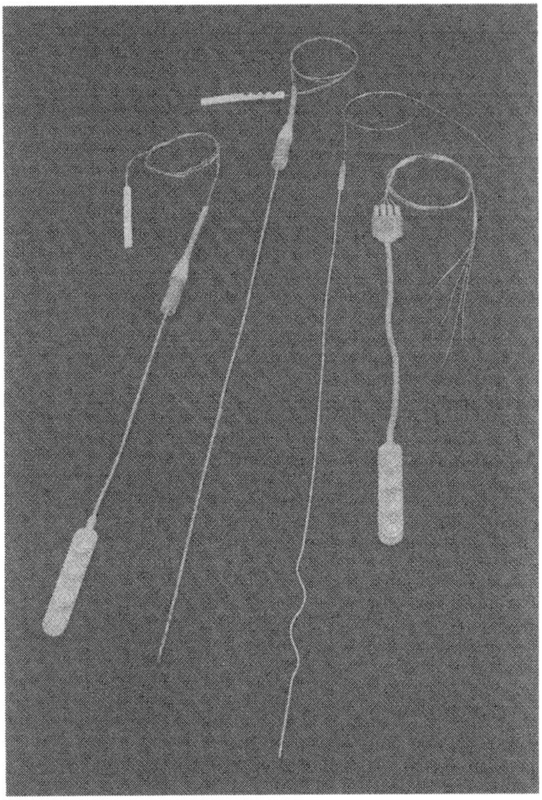

Fig. 4. Different kinds of multiple contact electrodes (Medtronic). On the left and on the right, electrodes introduced through a small laminectomy. The ones on the left allows to change the desired contact through the skin. Left in the middle, multiple contact electrode for percutaneous introduction. Right in the middle, monopolar sigma-electrode

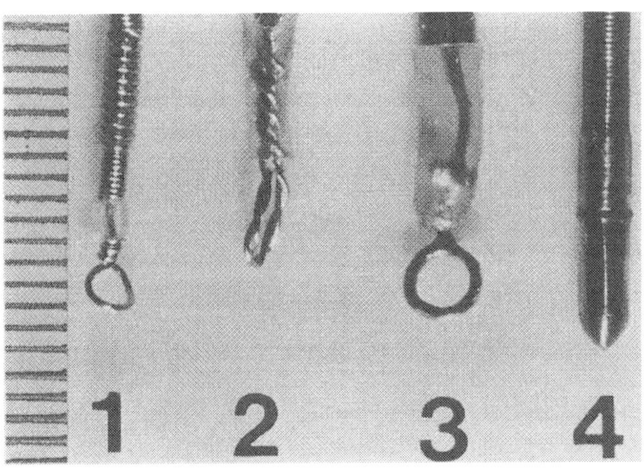

Fig. 5. Intracerebral electrodes. *1, 2* and *3:* Electrodes type to be used with a guiding cannula which engages into the loop (Shriver type). *4:* Avery electrode guided with a stylet in the center. (Measurement *(left side)* in mm)

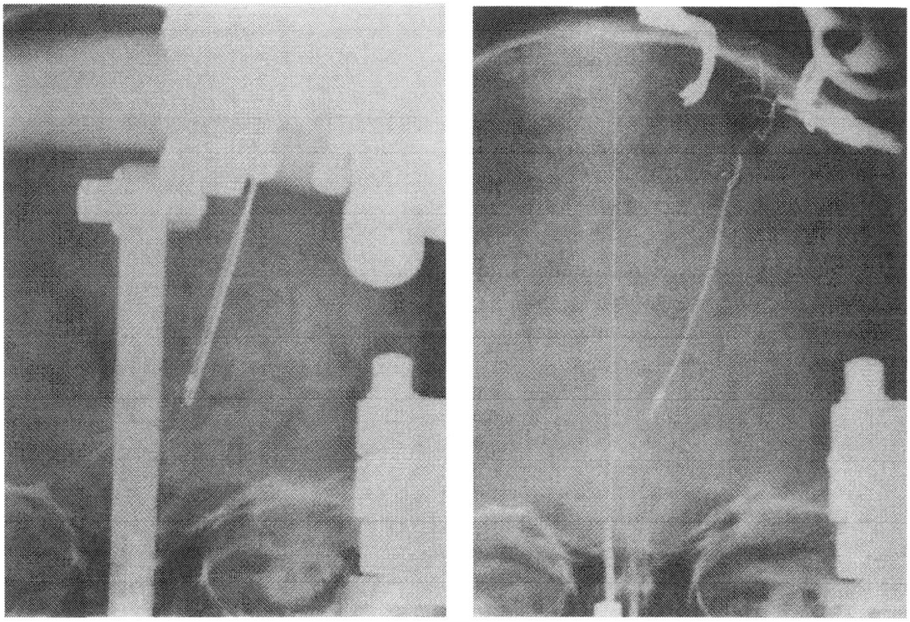

Fig. 6. Electrode migration within the brain immediately after removal of the stylet adjacent to it (Shriver type)

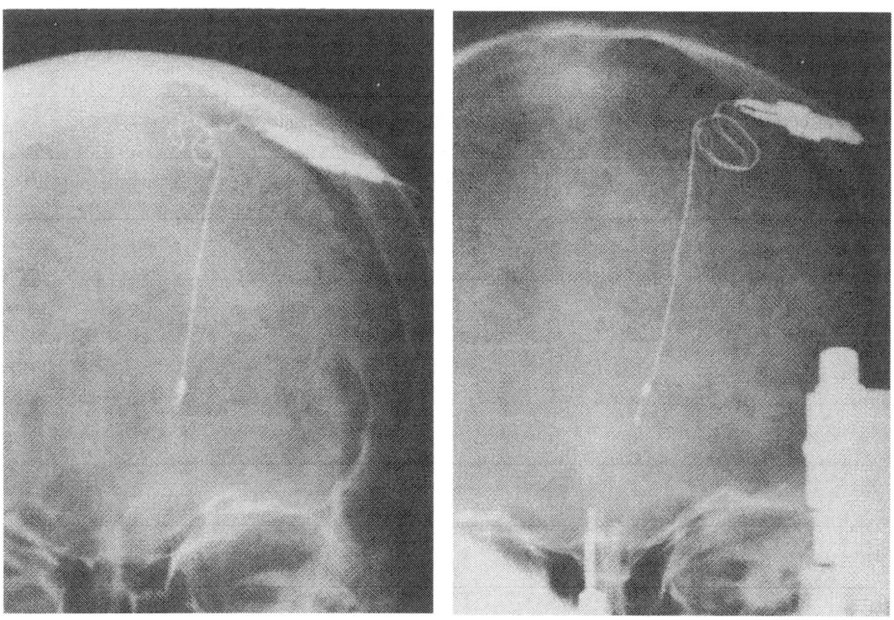

Fig. 7. Electrode migration within the brain 4 months after its placement (Shriver type)

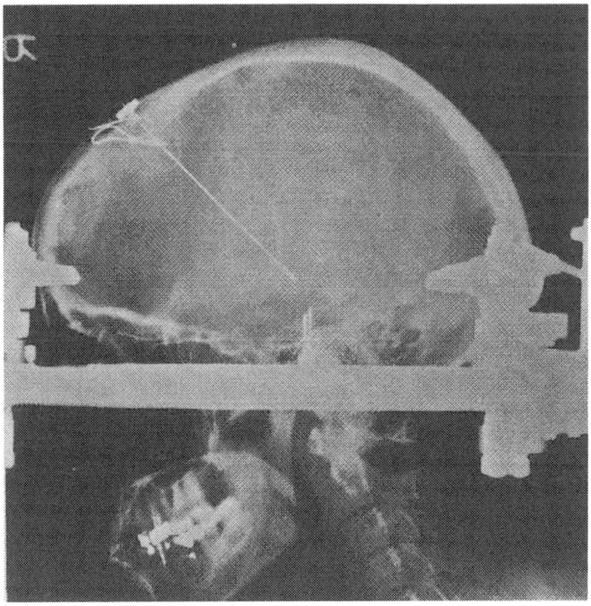

Fig. 8. Avery monopolar electrode fixed in the 2.5 mm burr hole with a Titanium screw

Intracerebral Electrodes

Electrodes within the cranial cavity may be used for either stimulation or recording, but have essentially the same basic requirements as the spinal axis electrodes (Comte et al. 1983; Maxwell et al. 1983; Siegfried et al. 1983). They are also made of nontoxic material such as platinum or stainless steel and cause little or no reaction with the CNS tissue. Because of the risks inherent in intracranial electrode placement, prevention of electrode migration and ease of placement are of paramount importance (Adams and Hosobuchi 1977; Miles and Dervin 1982; Richardson and Akil 1977). Some of the early electrodes were difficult to place due to their stylet being fixed parallel to the electrode (Fig. 5). Others had the tendency to retract along their path of insertion when the stylet was withdrawn and had no adequate fixation apparatus to the skull (Fig. 6, 7). Either of these deficiencies can lead to problems with electrode movement from the target site. An electrode which begins to overcome these problems has been manufactured by Avery (Siegfried et al. 1983). This electrode is hollow, providing for a central stylet. It is helical in design with platinum-iridium wire, making it flexible at the skull entry site, and at the same time giving it virtually no resiliance and no tendency to withdraw its insertion tract. It can also be fixed to the skull with a specially designed watertight titanium screw (Fig. 8). This system has been used in 111 consecutive patients at the University of Zurich in the past 5 years.

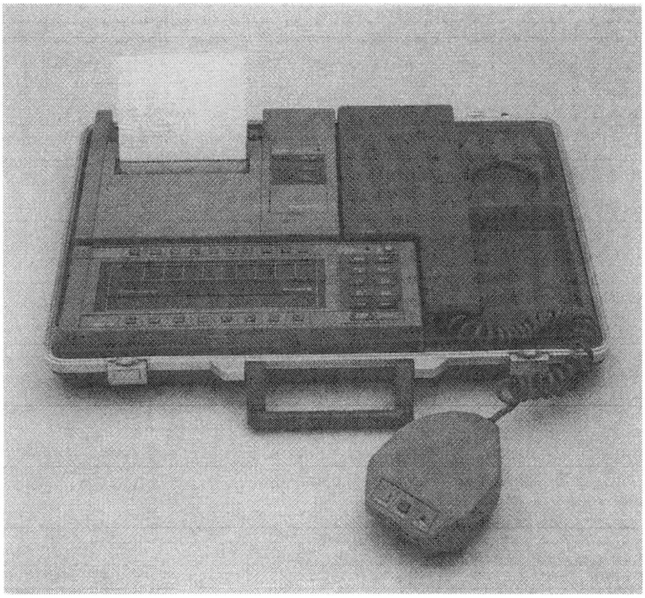

Fig. 9. Programmer for the ITREL totally implantable spinal cord or deep brain stimulation. Stimulation rate, amplitude, pulse width and length of on period of stimulation and off period between stimulation can be programmed transcutaneously and modified any time

Stimulator

Because the stimulators used in deep brain areas and the spinal axis are the same, the developments and improvements in the stimulating apparatus can be discussed in one section, without regard to the actual site of stimulation.

There have been two broad and closely related areas of stimulator improvement in the past decade, each of which has required multiple supporting alterations in order to reach fruition. These two areas are in programability and internalization (Delgado 1977; Mundinger et al. 1982; Seligman 1982). With the early stimulators there was little modifiability of the stimulus characteristics, but in recent models the amplitude, the pulse width, the timing of stimulation, the rise time of the pulse, and many other stimulus attributes can be altered (Seligman 1982).The capability to vary these parameters non-invasively (the stimulator's programability) increases the potential patterns of stimulation, and therefore increases the likelihood that the physician and the patient can find a therapeutic combination of stimulation characteristics.

The second significant improvement in stimulator design is the capability to totally internalize the unit (see Fig. 2). There are several advantages in this as compared to the use of a transcutaneously coupled system. These include: (1) the patient does not have a reminder of the pain necessitated by applying the external apparatus, (2) there are no external devices to break, and (3) the physician can have complete control of the parameters used. In this way the need for alterations, the lack of effectivness, or the desire to have the stimulation turned off, are all conveyed to the physician (Fig. 9). This is especially important to clinicians involved in closely monitoring the results of these devices.

At the present time, we have implanted 64 programmable and totally implantable systems for spinal and deep brain stimulation.

In summary, there have been significant improvements in electrode and stimulator design in neurostimulating devices over the past 10 years. Electrodes can be placed percutaneously with physiological and radiological confirmation of position, and the stimulator can be programmed in a multitude of different combinations after it is subcutaneously placed. Although these technological advancements are impressive, they have not eradicated the problem of chronic pain, and these devices continue to require careful clinical judgement in their use and close objective scrutiny of their efficacy.

References

Adams JE, Hosobuchi Y (1977) Technique and Technical Problems. Neurosurgery 1:196–199

Avery R, Comte P, Haut H, Siegfried J (1980) Experimental study of a Dacron Stabilized Electrode for Epidural Spinal Cord Stimulation. Proc Europ Soc for Artifical Organs VII:62–66

Blume H, Richardson R, Rojas C (1982) Epidural Stimulation of the Lower Spinal Cord and Cauda Equina for the Relief of Intractable Pain. In: Failed Low Back Surgery. Appl Neurophysiol 45: 456–460

Brummer SB, McHardy J (1977) Current Problems in Electrode development. In: Hambrecht FT, Reswick JB (eds) Functional Electrical Stimulation. Dekker, New York, pp 499–514

Brummer SB, Robblec LS, Hambrecht FT (1983) Criteria for Selecting Electrodes for Electrical Stimulation: Theoretical and Practical Considerations. Ann NY Acad Science 405:159–171

Cobun B (1980) Electrical Stimulation of the Spinal Cord: Two-Dimensional Finite Element Analysis with Particular Reference to Epidural Electrodes. Med and Biol Eng and Comput 18:573–584

Comte P (1982) Monopolar Versus Bipolar Stimulation. Appl Neurophysiol 45:156–159

Comte P, Siegfried J (1979) Therapeutical Spinal Cord Stimulation: Comparison between Monopolar and Bipolar Systems. Proc European Soc for Art Organs VI:345–349

Comte P, Siegfried J, Wieser HG (1983) Multipolar Hollow-Core Electrode for Brain Recordings. Appl Neurophysiol 46:41–46

Delgado JMR (1977) Instrumentation, Working Hypothesis and Clinical Aspects of Neurostimulation. Neurosurgery 1:191–195

Donaldson PEK, Donaldson NN, Brindley GS (1985) Life of P^+ and P^+-Ir Stimulating Electrodes in Neurological Protheses. Med Biol Eng Comput 23:83–84

Dymond AM, Knechelle LE, Jurist JM, Candell PH (1970) Brain Tissue Reactions to some Chronically Implanted Metals. Journal of Neurosurgery 33:574–580

Law JD, Muller LV (1982) Importance and Documentation of Epidural Stimulating Position. Appl Neurophysiol 45:461–464

Martinez N (1985) Personal Communication

Maxwell RE, Gates JR, Fial ME, Johnson MJ, Yap JC, Leppik LE, Gummit RJ (1983) Clinical Evaluation of a Depth Electroencephalography Electrode. Neurosurgery 12:561–564

Miles J, Dervin T (1982) An Electrode for Prolonged Stimulation of the Brain. Appl Neurophysiol 45:449–455

Mortimer JT, Shealy CN, Wheeler C (1970) Experimental non-destructive Electrical Stimulation of the Brain and Spinal Cord. Journal of Neurosurgery 32:553–559

Mundinger F, Neumuller H (1982) Programmed Stimulation for Control of Chronic Pain and Motor Disease. Appl Neurophysiol 45:102–111

Pudenz RH (1977) Adverse Effects of Electrical Energy Applied to the Nervous System. Neurosurgery 1:190–191

Richardson DE, Akil H (1977) Long Term Results of Periventricular Gray Self-Stimulation. Neurosurgery 1:199–202

Seligman LJ (1982) Physiological Stimulators: From Electric Fish to Programmable Implants. IEEE Trans Biomed Eng 29(4):270–284

Sharkey PC, Dimitrijevic MM, Fagenel J (1982) Neurophysiological Analysis of Factors Influencing Efficacy of Spinal Cord Stimulation. Appl Neurophysiol 45:68–72

Siegfried J, Comte P, Meier R (1983) Intracerebral Electrode Implantation System: Technical Note. Journal of Neurosurgery 59:356–359

Advances in Drug Delivery Systems: Neurosurgical Applications

R. D. Penn

The concept of a "chemode", a device to deliver medications to specific areas of the brain is an old one. Such a system would make control of neural function possible through the application of specific chemicals in a precise fashion to a limited region. This concept has now become a practical reality for use in man by the development of programmable implanted drug pumps. Coupled with catheter systems and stereotaxic placement, the neural environment can be changed in virtually any region. The initial neurosurgical application of this device took place three years ago, when a Medtronic pump was used to deliver morphine into the lumbar subarachnoid space to relieve cancer pain (Penn et al. 1984). The diffusion of morphine into the dorsal cord from the cerebral spinal fluid provided complete pain control without the undesirable central side effects. The dose could be precisely adjusted to the patient's needs.

Table 1. Drug pumps in neurosurgery

Use	Drug	Location	Status
Pain	Morphine	Spinal CSF	Proven
	Clonidine	Spinal CSF	In trials-human
	DADL	Spinal CSF	In trials-human
	Marcaine	Spinal CSF	Animal trials
	Marcaine	Epidural	Limited human trial
Spasticity	Baclofen	Spinal CSF	In trials-human
	Morphine	Spinal CSF	In trials-human
Alzheimer's	Bethanechol	Intra-ventricular	In trial-human
	Somatostatin	Intra-ventricular	Animal tox studies
	Physostigmine	Intra-ventricular	Animal tox studies
Gliomas	BCNU	Selective	In trials-human
	5-FU	Arterial	In trials-human
	Bleomycin	Cyst cavity	In trials-human
	CIS platin	Intratumoral	Animal
	5-FU	Intratumoral	Animal
Nerve block	Local anesthetics	Local	Animal
Epilepsy	NA valproate	Cerebral cortex	Animal
	Inhib aminoacids	Cerebral cortex	Animal

The treatment of cancer pain has been a good clinical area in which to test the new pumps, but many other applications are being actively investigated. A partial list of these studies is given in Table 1. In this article I will briefly describe the various pump systems, present the results in three different areas and then indicate the direction in which I think the field will develop.

Drug Delivery Systems

A wide variety of implantable drug delivery systems are under investigation. The simplest are ports for injections of drugs percutaneously into a reservoir leading to a catheter in the CNS. Limitations of these systems include inability to control with precision the volume delivered, the need for repeated injections, and the risk of infection. Because of these drawbacks, implanted mechanical pumps using CSF shunt technology are being designed and tested. A reservoir is connected to a metered valve which, when pressed by the patient, delivers medication. The reservoir is refilled percutaneously as needed via a port. The advantages of such systems are their simplicity, relatively low cost, and patient control. The disadvantages include only single bolus, fixed relatively high volume dosing and lack of limitation of the total dose.

A freon gas powered system (Infusaid, Inc.) which delivers a fixed volume per day has been available for several years, and a number of studies have demonstrated its reliability (Buchwald et al. 1980; Onofrio et al. 1981; Greenberg et al. 1982). The dose can be changed only by replacing the fluid in the reservoir with medication at another concentration. It cannot be turned on and off. Furthermore, changes in pressure (altitude) and body temperature can increase or decrease the dose by 20%–30%. For applications which require great precision, this system is not appropriate, but for an application such as the delivery of intrathecal morphine, in which a wide dosage range is tolerated well, it is an excellent system.

The programmable electronically controlled pumps which are currently being tested have a number of advantages over the other systems. A list of characteristics of the most widely used device made by Medtronic, Inc., which is given in Table 2 and Fig. 1, illustrates the various modes of delivery which can be programmed. Because precise microliter quantities of medication can be given, a small volume of concentrated medication can last for many months. The advanced pacemaker technol-

Table 2. Features of the implantable device

Medium volume reservoir – 20 ml
Programmable via telemetry
Can be off and on
Variable rates of delivery
Easy access for refilling
Size – 70 mm × 28 mm

Infusion Mode Name	Drug is Dispensed	Graphic Representation of Infusion Mode
1. Bolus	once, at the pump's maximum dispense rate	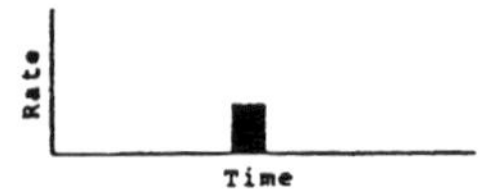
2. Multi-step bolus	once, in a series of 1-10 doses at a specified rate. The example shows a four-step bolus	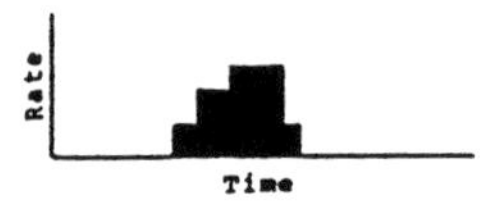
3. Continuous	continuously, at a specified hourly rate	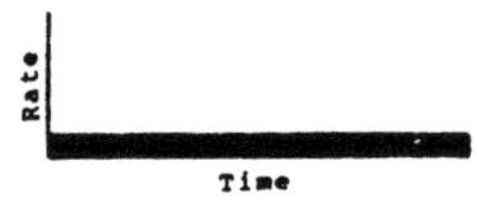
4. Continuous-complex	continuously, in a series of 1-10 doses at specified intervals. The example shows a five-step dose	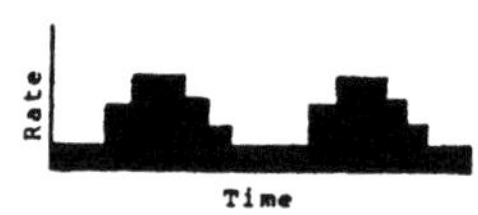
5. Bolus-delay	intermittently, at a prescribed dose and at specified intervals	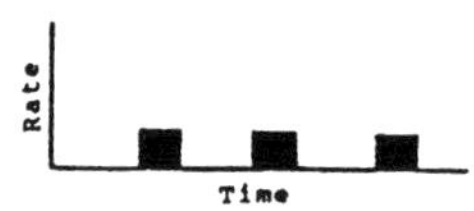

Fig. 1. Graphic Representation of Infusion Modes Employed by the Medtronic Model 8610 and Model 8611 Drug Administration Devices

ogy is evident in the design of the device which uses computer clips, RF telemetry, and a lithium battery to drive a rotary pump. As might be anticipated with such a complex system, a number of failures occurred with the first model. All of these were in a safe failure mode, the device simply stopped working. No medication was inadvertently given and no leaks from the pump reservoir into the tissue occurred. A new design has now overcome these problems and over 90 new systems have been implanted and are working properly. This achievement is remarkable in view of the need to blend pacemaker and pump systems into an integrated design. Currently, the computer used to control the device is bulky and only suitable for office, hospital, or operating room use. However, a hand-held device will soon be available, which will make possible patient control of dose within prescribed ranges if that is important for a particular application.

Table 3. Pain relief using spinal morphine for cancer pain

	Excellent	Good	Fair	Poor	Total
Programmable	7	2	1	1	11
Continuous	8	5	1	1	15
					26

Pain

The observation that severe cancer pain could be relieved for many hours by lumbar intrathecal morphine was the impetus for the use of systems to chronically administer the drug (Wang et al. 1979). The advantages of inhibiting pain on a spinal level are many, including no change in mentality, no respiratory depression, and no constipation. Tolerance does develop in some patients, but less than might have been expected, and it can often be handled by increasing morphine doses (Penn et al. 1984; Greenberg et al. 1982). A number of centers have used implanted systems to deliver morphine, either epidurally or intrathecally. Our results in 26 patients using the Medtronic and Infusaid systems are representative of the overall clinical experience (see Table 3). It should be emphasized that patients are only considered for implantation if oral narcotics have failed and if the patient has had relief with a trial dose of intrathecal or epidural morphine. The two poor results occurred early in the series when the need for a trial period of epidural morphine was not fully appreciated. The overall results are encouraging. Since percutaneous epidural catheters have also been shown effective, the implantation of drug pumps will be reserved primarily for patients with a long life expectancy.

The success demonstrates dramatically how regional perfusion of an area of the nervous system can be used effectively to block selected neural circuits. Of considerable interest in the future are other drugs which act on the pain pathways, such as other opiate receptor groups, and descending inhibitory mechanisms. Neural modulatory peptides, such as somatostatin, may also be useful. The ability to directly infuse the CSF means that problems of the blood-brain barrier can be avoided, so many medications which are not useful orally may be helpful if given intrathecally. Thus a whole range of medications for spinal pain control may be available as pharmacological advances are made.

Spasticity

The same concept of avoiding central side effects while maximizing spinal inhibition of neural circuits can be applied to spasticity. The most effective oral drug, baclofen, is limited by the drowsiness or confusion it produces. The same drug given intrathecally in the lumbar space, in microgram quantities can completely stop spasms and decrease muscle tone to normal in spastic patients with spinal cord damage

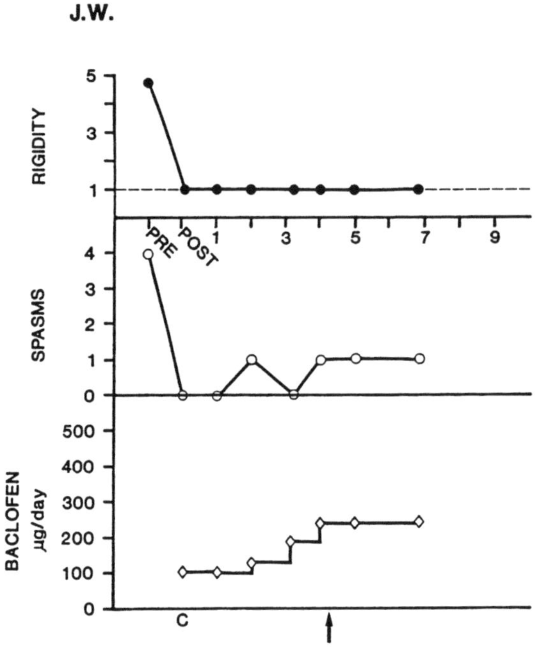

Fig. 2. Response of a 42-year-old paraplegic due to spinal cord injury to chronic intrathecal baclofen given via a Medtronic implanted drug pump. The severe spasms (> 10 per h) and board-like rigidity (5 on scale) have been completely controlled. The daily dose of baclofen needed to achieve this response increased for several months and is now stable

(Penn and Kroin 1984). This observation has led to the chronic infusion of spastic patients using an implanted drug delivery system. Since precise dosage is necessary, the Medtronic unit has been employed.

An example of one spinal cord injury patient followed for 10 months is given in Fig. 2. The patient's spontaneous spasms, which were greater than 10 per h (5 on the scale) were stopped, and her board-like rigidity (5 on scale) was reduced to normal. The baclofen dose had to be increased gradually for the first few months, but has now stabilized at just below 300 µg/day. No central side-effects have been seen in this patient in whom 30 mg of oral baclofen had produced mental confusion.

Results so far have been encouraging enough in these patients to organize a multicenter trial. Long-term observations will be necessary to determine the value of the method, since the patients have a near-normal life expectancy. Certainly it is more appealing to give medications than use destructive neurosurgical or chemical procedures in these severely affected patients.

Intratumoral Chemotherapy

The ability to place probes into tumors using CT or MNR stereotaxy is a subject of several papers in this symposium. The question is how this ability can be used to effectively treat the lesions. Drug infusion directly into a brain tumor is one possibility, and we have been investigating this experimentally in animals for several years. The key to delivery of medications into tissue is chronic low volume perfusion. Large,

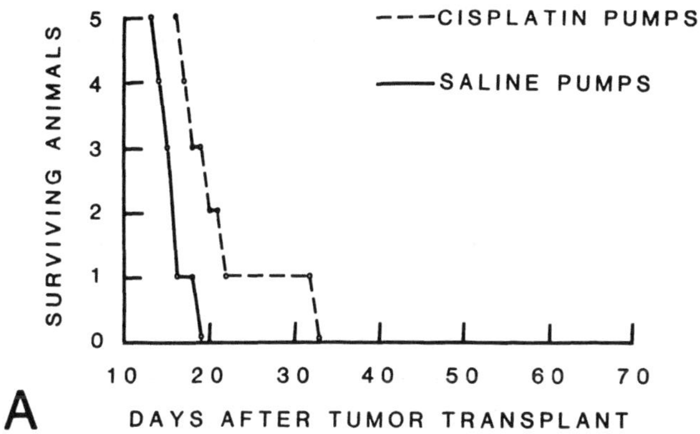

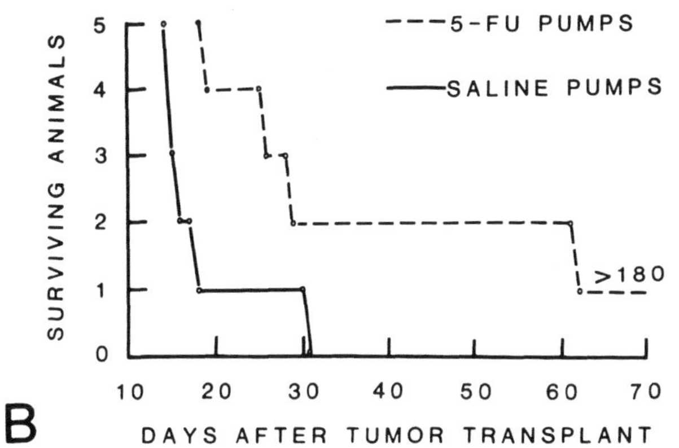

Fig. 3A, B. Survival of rats injected with 1×10^5 9L brain tumor cells in the cerebellum. Intratumoral chemotherapy commenced on day 8 following the tumor transplantation and lasted 7 days. The drug was infused at a rate of 0.9 μl/h and contained cisplatin, 0.5 mg/ml (**A**), or 5-FU, 25 mg/ml (**B**). Control animals had isotonic saline infused at the same rate

single dose injections are bound to fail because of tissue resistance and subsequent fluid loss along the needle or catheter tract. Slow small doses are spread by diffusion and therefore will be distributed around a catheter tip. Studies perfusing cisplatin into normal brain showed the effective distance of diffusion is small, about a centimeter in diameter (Kroin and Penn 1982).

In the animal model, with a small tumor, we have shown that a one week infusion of 5-FU or cisplatin can increase longevity significantly (see Fig. 3). Thus, these drugs, which have no effect systemically, can work if introduced directly into the growing tumor.

The difficulty with the technique is that human tumors are much larger, so multiple sites of delivery are needed. A point source is necessary for approximately every cubic centimeter of the tumor and surrounding region which must be treated. The technical problems of multiple catheters have yet to be fully solved, but with effort it may be possible to develop such systems. The important point is that the stereotaxic systems now exist for multiple source interstitial radiation in humans, and these could be adapted to catheters. It would be naive to think that perfusion of a tumor with the present drugs would be sufficient to cure gliomas. However, many chemotherapeutic agents, which have not been of value because of blood-brain barrier delivery problems could be considered. Also, other agents that might be immunologically active should be considered.

The Future

The interstitial application of chemicals into selected brain areas may eventually be the most important neurosurgical use of programmable implanted drug pumps. While the small distance that drugs diffuse in the brain may make treatment of tumors and generalized brain disorders difficult, the limits of diffusion make possible very precise targeting of chemotherapy. Using neurotransmitters, or neuromodulators to influence the behavior of small groups of neurons could well replace ablative stereotaxic procedures. It should also make possible new applications, since these agents will effect selective cell populations. Now that the tools are available, the challenge of the next decades will be to use it properly to treat a wide range of neurological disorders.

References

Buchwald H, Rohde TD, Dorman FD, et al (1980) A totally implantable drug device: laboratory and clinical experience using a model with single flow rate and new design for modulated insulin infusion. Diabetes Care 3:351–358

Greenberg HS, Taren J, Ensminger WD, et al (1982) Benefit from and tolerance to continuous intrathecal infusion of morphine for intractable cancer pain. J Neurosurg 57:360–364

Kroin JS, Penn RD (1982) Intracerebral chemotherapy: chronic microinfusion of cisplatin. Neurosurgery 10:349–354

Onofrio BM, Yaksh TL, Arnold PG (1981) Continuous low-dose intrathecal morphine administration in the treatment of chronic pain of malignant origin. Mayo Clin Proc 56:516–520

Penn RD, Kroin JS (1984) Intrathecal baclofen alleviates spinal cord spasticity. Lancet I (8385): 1078

Penn RD, Paice JA, Gottschalk W, et al (1984) Cancer pain relief using chronic morphine infusion: early experience with a programmable implanted drug pump. J Neurosurg 61:302–306

Wang JK, Nauss LA, Thomas JE (1979) Pain relief by intrathecally applied morphine in man. Anesthesiology 50:149–151

Implantable Systems for Local Chronic Administration of Drug Applications in Neuropharmacology

Y. LAZORTHES and J.-C. VERDIE

Recent neurobiological data concerning the existence of multiple receptor sites and the individualization of specific neurotransmitters have forced us to step back and take another look at some of our preconceived ideas about the central nervous system. The concept of a chemical neurotransmission, complementary and specific to a given transmission, has progressively been associated with the electric model.

A better knowledge of the action mechanism of the endogenous and exogenous ligands has allowed us to dispose numerous active compounds. The systemic approach to drug administration requires large doses that eventually reach the action site, but only after having selectively crossed the blood-brain barrier. In addition, the quantity of drugs administered causes a variety of undesirable side effects. The difference between the dosage which is administered systemically and the dosage which is actually therapeutically effective is the basis of development of a new therapeutical approach: loco-regional neuro-pharmacology. This method has progressively developed at several sites of administration (epidural, intrathecal, intraventricular, intravascular, intraparenchymally, etc.) for numerous drugs (opiates, anti-mitotics, antispastics, neuromediators, etc.) and for a variety of indications.

The need for chronic administration of drugs either through continuous perfusion or by repeated injections led to the development of implantable systems, permitting us to:

- increase the comfort and quality of life of those treated on an out-patient basis
- develop the concept of "pharmacological neurosurgery" which appears to be a method of the future as it is not only effective, conservative, and selective, but also completely reversible
- and finally, to optimize the therapeutic effects of chronotherapy.

Material

The implantable systems are connected to supple silicone catheters whose distal ends are positioned near the particular action site. There are two types of systems.

Reservoirs

These consist of small-volume injection sites which are shunted to the area to be perfused. The goal of these "access sites" is to transform a lumbar or an intraventricular

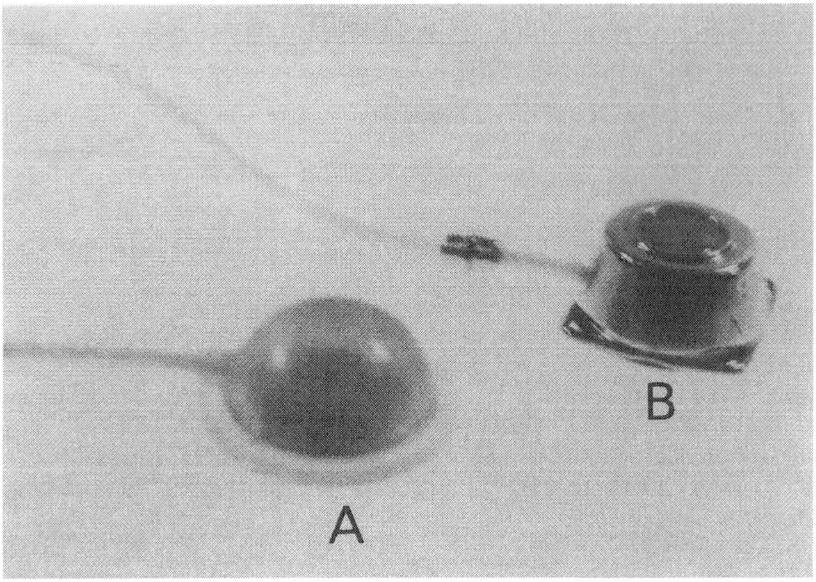

Fig. 1 A, B. The two models of implantable reservoirs. **A** The Cordis unidose reservoir. **B** The Pharmacia PORT-A-CATH system

puncture or an intra-arterial injection into a simple transcutaneous injection. There are two models (Fig. 1):

1. Silicone, dome-shaped capsules originating from the Ommaya reservoir but with a flat bottom and side outlet. The CORDIS[1] "unidose" reservoir is the most widely known; it consists of a stainless steel reinforced bottom to prevent perforation while injecting. These small-volume injector-reservoirs (1–2 ml volume) correspond to a single dose injection.

2. Stainless steel injection sites with a self-sealing chamber. This is the PORT-A-CATH system manufactures by Pharmacia[2]; it can be connected to an intravenous or intra-arterial catheter as well as to a catheter to be placed epidurally or in the subarachnoid space.

These implantable reservoirs allow drugs to be administered either by repeated injections or by slow continuous perfusion (a maximum rate of 1 l/h). They provide no autonomy to the patient who must depend on a prescription "on demand". It is always difficult to be sure you are injecting the entire contents of the reservoir, and unfortunately, none of these systems include an integral anti-bacterial filter.

In order to provide his patients with a certain autonomy, Poletti (Poletti et al. 1981) devised a system whereby a lumbar catheter was connected to a PCB-500 cc silastic-coated blood pack which was implanted in the lining of the abdomen. He incorporated a high-pressure Hakim valve and an on-off valve, permitting the transcutaneous injection of a 0.1 cc bolus. This combination of pre-existing parts represents an original, economical solution; however, it is bulky and not very functional.

[1] Cordis Europa, PO Box 38, AA 9300 Roden, The Netherlands.
[2] Pharmacia France SA, 8 square Newton, 78391 Bois D'Arcy.

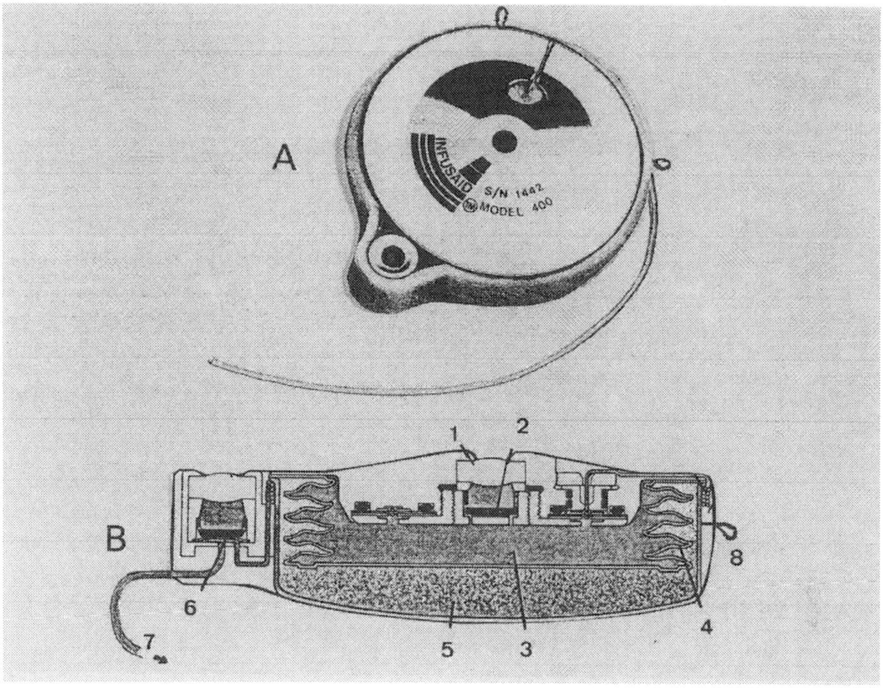

Fig. 2 A, B. The constant-flow implantable pump of Infusaid. **A** The pump model 400. **B** A cross section – *1:* inlet septum, *2:* needle stop, *3:* drug chamber (50 ml), *4:* titanium metal bellows, *5:* charging fluid chamber (fluoro-carbon), *6:* auxillary septum, *7:* outlet catheter, *8:* suture loop

Implantable Micro-Pumps

These pumps have a large enough reservoir capacity to allow patient autonomy and they can easily be refilled through simple percutaneous injection. The different systems currently available vary in their working mechanism and in their capability to insure safety and provide accurate, variable prescriptions.

1. If we consider the working mechanisms, we can differentiate between the following:

– Pumps whose means of propulsion is chemical, in the form of Fluorocarbon gaseous pressure (Freon). In 1969, B. J. Blackshear (1979) invented the system which is manufactured by Infusaid[1]. The disc-shaped pump is 3 cm high and has a diameter of 9 cm; its empty weight is 200 grams and it has a reservoir capacity of 50 ml (Fig. 2). The pump is separated into 2 chambers by titanium bellows. The inner drug chamber contains the solution to be infused and the outer charging chamber contains a fluorocarbon liquid in equilibrium with its vapor phase. At 37°C, the vapor pressure exerted by this substance is approx. 300 mmHg greater than atmospheric pressure. This vapor pressure provides the power source, exerting pressure on the bellows and

[1] Infusaid Corp., 1400 Providence Highway, Norwood MA 02620 USA.

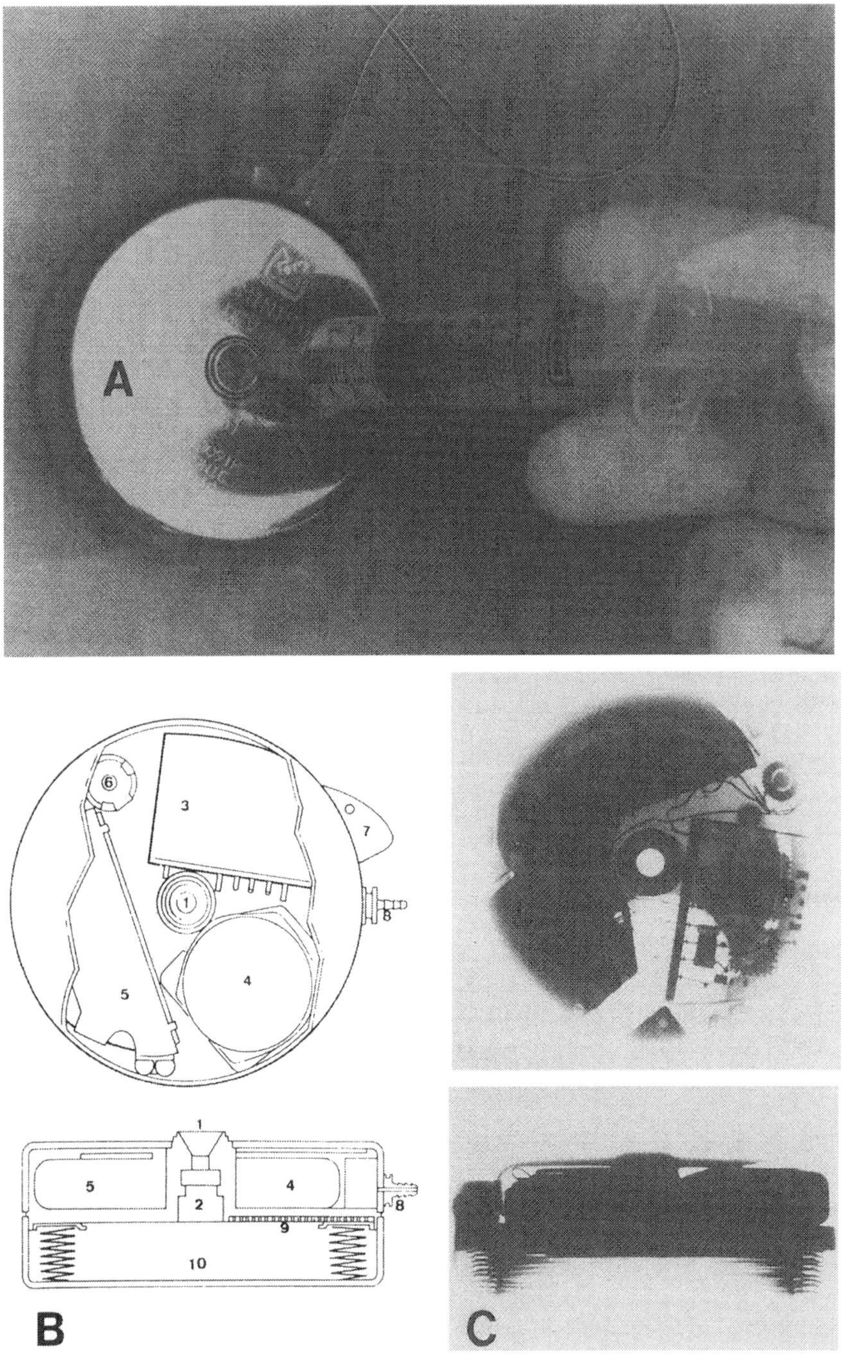

Fig. 3 A–C. The medtronic implantable and programmable micropump: The drug delivery device "D.A.S.". **A** Photography of the system, model 8600. **B** Diagramatic section – *1:* fill port and self-sealing septum, *2:* needle stop, *3:* electronic module, *4:* peristaltic pump, *5:* lithium battery, *6:* antenna. *7:* suture pad, *8:* titanium connector, *9:* bacterial retentive filter, *10:* collapsable reservoir (20 ml). **C** Front in view and profile X-ray of the device

forcing the infusate through a 0.22 μm bacterial filter and flow-regulating resistance element. The pump is placed beneath the skin, and the inner drug chamber is refilled by percutaneous injection. The pressure of this injection condenses the driving vapor, thus simultaneously filling the pump and recharging the power source.

– The electro-mechanical peristaltic pumps are the most well known. The peristaltic pump consists of a flexible tube placed in a metallic U-shaped chamber. Roller squeeze the tubing walls at variable speeds which propels the drug at a variable flow-rate. The most sophisticated model is the one manufactured by Medtronic Inc[1]: The Drug Administration System or D.A.S. 8600 model (Fig. 3). The reservoir, drug pathways and exterior of the D.A.S. are constructed of titanium elastomers; it measures 2.75 cm high and has a diameter of 7 cm. Its empty weight is 175 grams with a reservoir capacity of 20 ml. The 8601 model includes a bacterial retentive filter of 22 μm. Energy is provided by a lithium thionyl-chloride battery with a life expectancy of approximately 10 years. The programmable infusion parameters are controlled by a microprocessor which is hermetically sealed in the leak-proof D.A.S. case. The pump made by Siemens (IDE Model) is similar (Table 1).

– The electro-mechanical pulsatile solenoid pumps, developed by Pacesetter[2], uses a chamber which is activated by an alternate coming and going movement produced by the solenoid. Two feed-check valves move the drug out towards the catheter. Energy is provided by a lithium battery. The pump measures 2 cm high and has a diameter of 8.1 cm. The empty weight is 170 g, and it has a reservoir capacity of 7.2 ml. The system is controlled by a microprocessor-based circuitry which permits programmation and interrogation by telemetry. This system has only been used as a prototype.

– The totally mechanical pumps. The "SECOR" pump designed by Cordis represents the first implantable pump activated by a simple manual gesture (Lazorthes et al. 1985). It is entirely mechanical, including no electronic features nor battery; thus it is not a programmable system. The working mechanism consists of 3 one-way valves which control and deliver doses of a constant volume. The disk-shape pump has a diameter of 6 cm and a height of 1.4 cm. Its empty weight is 45 g (Fig. 4). On the top of the pump you can see 2 push-buttons which are activated transcutaneously. They must be pressed sequentially and in the correct order before a dose is delivered. A center self-sealing refill dome made of silicone elastomer permits percutaneous emptying and refilling. It includes a safety valve and a needle stop which prevents the needle from completely passing through the dome and entering the reservoir. The bottom of the pump includes a supple reservoir composed of teflon-reinforced silicone with a working capacity of 12 ml–15 ml.

2. If we consider the various pumps with respect to security and prescriptions, we can distinguish between the following:

– Constantly-flow pumps Infusaid. Considering the pump is activated by a gaseous pressure, fixed according to a given temperature, the flow is relatively stable but precalibrated during manufacturing. Having been set between 2 mg/24 h and 6 mg/24 h, it is impossible to change the flow rate after implantation. In order to modify the prescription, you need to change the drug concentration. One other inconvenience

[1] Medtronic Inc., 3055 Old Highway, PO Box 53, Minneapolis, MN 55450, USA.

[2] Pacesetter Syst., 12884 Bradley Av., Sylmar, CA 91342, USA.

Table 1. The implantables drug delivery devices

Systems	Mechanisms	Dimensions (cm)	Empty weight (g)	Infusion modes	Reservoir volume (ml)	Alarm	Program-mable	Side port
1. Infusaid Mod. 400	Vapor pressure	3 × 9	208	Continuous 2–6 ml/24 h	47	No	No	Yes
2. Medtronic Mod. 8600	Peristaltic	2.7 × 7	175	Variable 0–0.9 ml/h	20	Yes	Yes	No
3. Siemens Mod. IDI	Peristaltic	8.5 × 6 × 2.2	170	Variable (24 steps)	10	Yes	Yes	No
4. Pacesetter Johns Hopkins	Solenoid pulsatile	8.1 × 2	170	Variable (2 μl steps)	7.5	Yes	Yes	No
5. Cordis Secor	Mechanic	6 × 1.4	45	Bolus (1 or 3 ml)	12	No	No	No

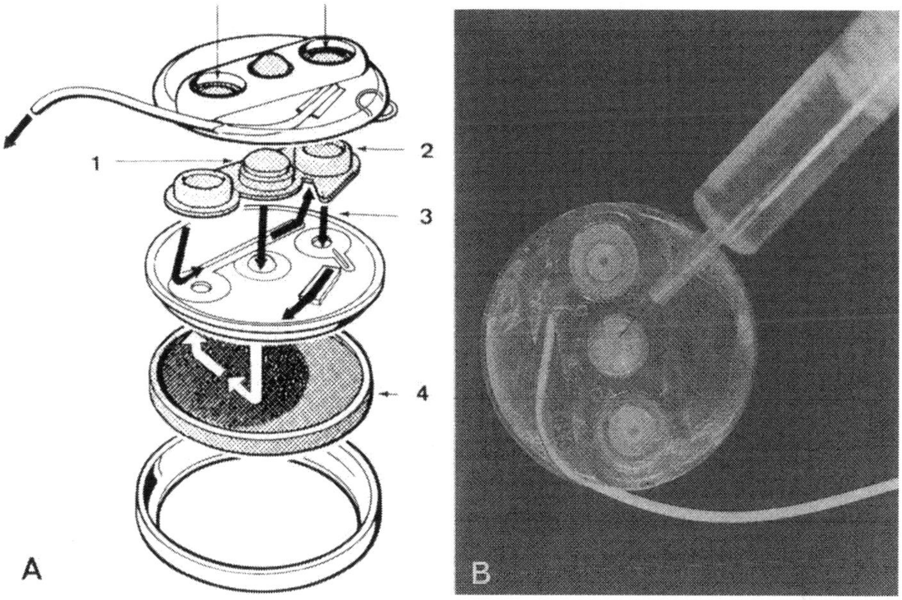

Fig. 4 A, B An implantable, totally mechanical multidose drug delivery system. The "SECOR" system of cordis. **A** Functional schedule – *1:* self-sealing refill dome, *2:* recessed push-buttons, *3:* one-way valve system, *4:* reservoir (12–15 ml). **B** Photography of the device

with this system is that the flow increases with changes in body temperature and diminishes with the atmospheric pressure. Moreover, it is impossible to immediately stop perfusion other than by emptying the drug reservoir percutaneously which entails a risk in the event of serious side effects. The 400 model includes a side outlet which can be used to administer another drug and to check the catheters permeability.

– The "multidose" pumps (Cordis "SECOR" model). The external manipulation of the 2 push-buttons releases a 0.1 ml (+ or – 10%) bolus. These buttons must be pressed sequentially and in a correct order before a dose is delivered so that they cannot be activated accidentally. To modify the flow, you simply increase or diminish the number of boluses, A new SECOR generation will have pop-up push-buttons and make a clicking sound to assure the manipulator that the bolus has been released.

– The multiprogrammable pump (Medtronic's D.A.S. model) (d) has 3 programmable infusion parameters that can be set or changed by telemetry: 1° a choice of measurements (μg, mg, ml, mmol, etc.), 2° a variety of infusion modes (continuous hourly infusion, infusion with a specified delay, multi-step dose over programmed interval, or bolus infusion) and 3° a range of infusion rates (approx. 0.09 ml/h to 0.9 ml/h) (Fig. 5).

Telemetry permits the clinician to non-invasively program the pump to a specific prescription, to control the D.A.S. operation and to interrogate the different alarm systems. In case of emergency, the programmer can be used to stop the pump. In order to do so, the patient must be transported to the hospital where the implantation took place. There exists portable programmers similar to those used for diabetics,

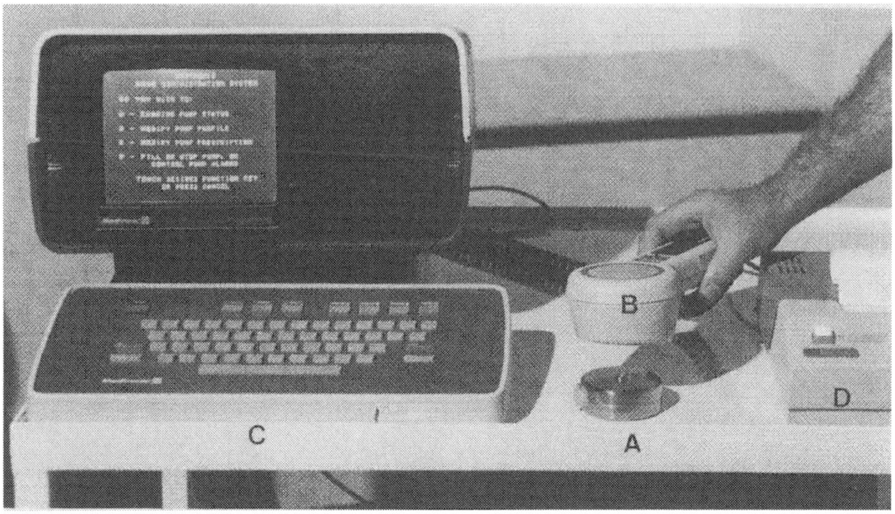

Fig. 5. The programmer of the "D.A.S." system. *A:* Implantable pump, *B:* programming wand, *C:* computer/screen unit and keyboard, *D:* printer

however, they are expensive and not everyone can afford them. To overcome this handicap, we hope that a simplified version with only an on-off control will be developed soon. There are also auditory alarms that indicate the reservoir capacity, the lifespan of the battery, and an eventual electronic breakdown.

The principle characteristics of these different implantable systems are reviewed on Table 1.

Implantation Technique

The implantation is normally done percutaneously under local anesthetics or neuroleptanalgesia (Lazorthes et al. 1980; 1982; 1985; Lazorthes and Verdie 1985).

Catheter Placement

Catheter placement depends on the drug administration action site: spinal, intraventricular or intra-arterial. The catheter must be passed under radioscopic-control to determine its exact location.

We use a radio-opaque silicone catheter with an outer diameter of 1.6 mm and an inner diameter of 0.75 mm (Cordis) for intraspinal and intraventricular placement.

A 14 G Touhy needle permitted us to place the intra-spinal catheter into the subarachnoid or epidural space. We used a percutaneous lumbar latero-median approach so that the puncture pathway was as oblique as possible in order to avoid catheter linkage during interlaminar penetration of the vertebral canal. The pump connection can either be made at the site of the paravertebral incision or close to the subcutaneous pocket which will house the reservoir. It is extremely important that the connection be leak-free to prevent CSF fistulas and subcutaneous drug leakage.

In only 3 cases was the subarachnoid catheter inserted under visual control during a laminectomy and securely sutured to the dura mater.

The intraventricular method was done via a classical ventricular puncture. The distal tip of the catheter is either placed in the frontal horn of the right lateral ventricle opposite the foreman of Monro or in the third ventricle.

When using an implantable system for intra-arterial chemotherapy treatment, we prefer a radio-opaque silicone catheter whose outer diameter is 2.29 mm and whose inner diameter is 0.72 mm. The distal end is equipped with a suture ring which permits catheter fixation (Pharmacia). This extremity is placed in the external carotid artery and catheterized going backwards to the carotid bifurcation.

Implantation of a Drug Administration System

Regardless of whether you are using a reservoir or a micropump you must make a subcutaneous pocket to house the system. Make sure that the pocket is situated 2 cm–3 cm beyond the incision not directly on top of it, but yet close enough to facilitate fixation and to control the catheter connection. The subcutaneous surface must be thick enough to prevent skin necrosis especially with large-size pumps; but on the other hand, the system must not be implanted too deeply in order to facilitate transcutaneous palpation. Each system has its own particularities.

– The reservoir-injector "Unidose" by Cordis for intraspinal and intraventricular drug administration. To simplify injection procedures, we normally place the unidose on a hard surface: either the external iliac fossa or on the cranial cavity nearest the action site. The entire system (catheter-connector-reservoir) should be straight as possible in order to avoid kinks and assure optimal permeability. Before fixing the reservoir in the subcutaneous pocket, pump it to test the permeability. The reservoir puncture technique is identical to the lumbar puncture: a 25 G or smaller Tuohy needle will preserve the self-sealing properties of the silicone dome. During drug administration, we can test the permeability of the system by observing a CSF reflux and take CSF samples for bacteriological analysis. Drug injection must be done slowly; once the needle is removed, the dome is pushed down so that the bolus is progressively ejected into the subarachnoid space. You can rinse the reservoir with an additional 1 ml–2 ml bolus of CSF which has been drawn prior to injection of the drug.

– The "PORT-A-CATH" intra-arterial system is placed in a subcutaneous subclavicular pocket. The chamber must be flushed with a heparin solution of 500 U/ml. After cutting the catheter at the desired length you connect it to the drug chamber. This chamber is then attached to the fascia with 4 nonabsorbable suture ligatures. It is important to test the permeability of the system by injecting a heparin solution using a 22 G needle. Perfusion can be done either with a syringe or with a portable pump (minimum flow rate of 5 cc/mn). To withdraw the needle, you must at the same time hold the chamber down between 2 fingers to avoid pulling out the suture stiches and continue injecting to avoid blood reflux at the catheter tip. An anticoagulant treatment is continued with daily injections of 3 ml–5 ml of heparin at 500 U/ml for the first 3 days. The 8th day, we make another injection then once every week until treatment is completely discontinued.

– The "SECOR" (Cordis) multidose reservoir. Its fairly large size requires implantation on the lower latero-thoracic level for intra-spinal administration, while the subclavicular level is more suitable for intraventricular administration. Before connecting the reservoir, you must purge it of all air bubbles by filling it with a sterile saline solution. The pump must be pre-operatively calibrated to check the bolus volume and filled with the drug solution. With a bolus volume of 0.1 ml (+ or – 10%) and a morphine concentration of 5 mg/ml, the bolus will correspond to 0.5 mg (+ or – 0.05) of morphine.

– The "Drug Administration System" (D.A.S. Medtronic). Prior to implantation, remove all water in the reservoir and fill it up with 20 cc of drug solution in a sterile manner. Using the programmer under sterile conditions, you can then proceed with the examination and programmation. You must feed in data concerning the patient, the metrology and the nature of the drug to be administered (morphine, antimitotics, saline solution, etc.). The limits of the "empty reservoir" alarm must then be set. In practice, according to the pumps specific flow rate, you must program the quantity of drug that can be consumed before the audible alarm rings. At this point, the pump is submerged in 37°C water to check if all the bubbles have been completely evacuated. Considering the pumps large size, it is preferable to make the subcutaneous pocket in the flank or in the para-umbilical abdominal area. The D.A.S. is now ready to be programmed. The programmer includes a computer-type console television screen, keyboard and transmission head. Programmation is simplified by the fact that the software guides the manipulator, eliminating any chance of error.

Thus, since 1978 we have implanted 62 "Unidose" reservoirs of which 46 were connected to intra-spinal catheters and 16 were connected to ventricular catheters. During the same period we implanted 5 multiprogrammable D.A.S. pumps and 6 multidose SECOR systems for clinical investigation purposes in conjunction with the French authorization committee.

Incidents and Complications due to Implantation

Catheter disconnection occurred in 3 cases. The diagnosis was made after observing the subsequent inefficiency of the injections and afterwards visualizing the leak through opacification. In each of the 3 cases, surgery repaired the connection and therapy continued normally. To prevent incidents of this sort, we now overlap the silicone catheter ends and make a ligature before suturing the metal connector to the fascia.

Two patients had local infections and the systems had to be removed. One case involved a cachectic patient (with multiple urinary tract and respiratory infections) who developed septic skin necrosis with the unidose system. Another case developed a late (3rd month) post-operative hematoma infection with a SECOR pump. This risk of infection imposes certain preventive measures: eliminate patients with underlying infection, insure a thorough subcutaneous hemostasis and select an implantation site where cutaneous and subcutaneous layers are adequate. Finally, strict aseptic precautions must be observed while puncturing, emptying and refilling the reservoir. Hence the value of pumps with a sufficient storage capacity, reducing manoeuvres of this type to a minimum.

Only one case of purulent meningitis was seen this being the most serious complication. In this case where it was not possible to isolate the germ, recovery occurred spontaneously by injecting antibiotics intrathecally via the reservoir. The reservoir remained in place and continued functionning normally for another 8 months.

With regard to implantable pumps we did not observe any specific complication. In the first SECOR multidose generation, it was difficult to transcutaneously palpate the push-buttons. This inconvenience was rectified with the second generation which includes a pop-up action. None of the 5 Medtronic (D.A.S.) programmable systems had any electronic failures and the flow-rate checks were always accurate.

Present Day Applications. Preliminary Results

Chronic Pain and Intrathecal Opioids

This method is based on neurobiological, experimental and clinical data:

– Opiate receptor sites have been discovered in the central nervous system and in particular in the most superficial layers of the dorsal horns of the spinal cord (Pert and Snyder 1973).

– In vertebrates, we have observed a direct spinal action mechanism after morphine had been injected intravenously (Conseiller et al. 1972) and after micro-iontophoretic injections of morphine in the gelatinous substance (Duggan et al. 1977).

– Morphine administered directly into the lumbar CSF in the rat produces a powerful, prolonged analgesia which is limited to the caudal extremity (Wang 1977). These effects suggest that specific opiate receptors are involved since they are dose-dependent, stereospecific and naloxone-reversible.

– An intensive, selective and prolonged (average 24 h) analgesic effect was observed in patients suffering from chronic pain who received small doses of morphine (0.5 mg–3 mg) injected both intrathecally (Wang 1985) and epidurally (Behar et al. 1979). The work of Yaksh and Rudy 61976) emphasize the considerable importance of this type of drug delivery both from a fundamental standpoint and its clinical applications.

This technique essentially applies to chronic pain secondary to inoperable cancer. The incapacitating pain which is not relieved by conservative medical techniques, in particular by administering oral opiates, generally has a midline bilateral or diffuse topography. Consequently, both open and percutaneous surgical procedures designed to interrupt the nociceptive pathways are ineffective.

If we compare the indications, we can distinguish 2 complementary drug administration sites:

a. Intrathecal lumbar administration of morphine is essentially recommended for pain with a sub-diaphragmatic topography. This well known technique has become a routine treatment in analgesic therapy for cancer patients. The terminal cancer pa-

Table 2. Spinal morphine for cancer pain

(Ref.)	Devices		Site		Dosis (mg/24 h)	Follow-up (months)
	R	P	EPI	I-TH		
Wang (1985)	–	4	–	4	2	–
Poletti et al. (1983)	2	–	2	–	2	6– 7
Onofrio et al. (1981)	–	8	–	8	1 – 4.8	3–34
Leavens et al. (1982)	2	–	–	2	1 – 4	4– 7
Saunders and Coombs (1983)	–	12	8	4	0.5–50	2– 8
Penn et al. (1984)	–	14	2	12	1.5– 8	1– 9
Lazorthes and Verdie (1985)	43	9	4	48	1 –10	1–14
					Mean = 2.5	Mean = 4

tient who has a life expectancy of only a few weeks can be effectively relieved with epidural morphine injections. The epidural catheter is tunnelized and connected to either a millipore filter or a portable external pump. The inconveniences of an external catheter for chronic pain relief are: the risks of catheter migration; a great risk of infection in an immuno-depressive population; and the numerous restraints that a strict supervision implies, above all on an out-patient basis. For these reasons, when the patients life expectancy is sufficient, the use of a totally implantable system is more suitable. Since 1978, we have implanted the following systems: 44 Unidose reservoirs, 6 Multidose "SECOR" pumps and 3 D.A.S. micropumps (1980, 1982, 1983). Table 2 is a summary of our experience, as well as that of other authors in this particular field of drug administration. We prefer using hyperosmotic preservative-free morphine. The daily dose is minimal (between 1 mg and 5 mg with an average of 2.3 mg per day). We had a patient follow-up of 1 month to 14 months with an average of 4 months. Clinical evaluation was based on three criteria: the degree of analgesia obtained using a subjective pain-rating scale; consequences concerning the patients functional activities; and analgesic consumption. All together, we had 75% positive results. Repeated bolus injections on demand seem to be a better choice than continuous intrathecal perfusion due to the variable evolving nature of pain. The most frequent side effect is transient nausea. We had 2 cases of respiratory depression accompanied by drowsiness and myosis; these were among our first patients and the complication occurred during the percutaneous pre-testing period after normoosmotic morphine administration. For this reason, we now prefer using a hyperosmotic morphine solution and require the patient to remain in a half seated position 2 h–4 h after the injection. Moderate tolerance developed in 6 of our patients but we did not need to interrupt treatment.

b. Intraventricular administration of morphine. More recently, different authors (Leavens et al. 1982; Roquefeuil et al. 1984) showed that intraventricular morphine injections in the lateral ventricle produced 12 h–52 h of pain relief with only 0.1 mg/h–2 mg/h of morphine in patients with pain of cephalic or diffuse topography due to ear-nose-throat cancer. Pain relief was not associated with neurological changes or central depression. In fact, very few cases of respiratory depression have

Table 3. Intraventricular morphine for cancer pain

(Ref.)	Systems		Dosis (mg/24 h)	Follow-up (days)
	R	P		
Roquefeuil et al. (1984)	8	–	0.2 – 7	8–120
Leavens et al. (1982)	4	–	0.5 – 7	2– 90
Lobato et al. (1983)	17	–	0.25–16	15–120
Nurchi (1984)	6	–	0.33– 4	–
Thiebaud et al. (1985)	32	–	0.12– 5	4–230
Lazorthes and Verdie (1985)	16	–	0.12– 1.5	12–140
			Mean = 0.66	Mean = 68

been reported with intraventricular injections. Euphoria and transient confusional hallucination syndromes are two side effects specific to this particular injection site. The activating mechanism is probably directly at the supraspinal level, but there is controversial discussion that it might complement action at the spinal level. The problem of tolerance did not interrupt treatment. Major publications are summarized in Table 3. Our present-day, experience with 14 patients corresponds with the other authors observations.

In practice, these two injection sites appear complementary in that lumbar administration is indicated for pain of a subdiaphragmatic topography, whereas intraventricular administration is effective and justified when pain is diffuse or located above the diaphragm. Other than being extremly effective, other advantages of this method are its conservative, scarcely invasive nature and the fact that when undesirable side effects do occur, they are immediately totally naloxone-reversible. Morphine which is a mu receptor agonist, remains the opiate of choice, but several studies are underway concerning the use of other opiates such as pentazocine, dynorphine or the enkephalines which are also receptor agonists. The future prospects of this approach are numerous and for the most part related to the development of the drug-release systems and their capability to adapt to the needs of the cancer patient. One of the limitations, we are facing is the high cost of such implantable systems. Repeated bolus injections on demand appear to be more effective and easier to control than continuous perfusion.

Spasticity Control Through Intrathecal Drug Administration

The applications in this field are more recent and the neurobiological, experimental and clinical basis is still somewhat limited:

– J. C. Willer demonstrated that intravenous morphine injections (0.2 mg/kg–0.3 mg/kg) in volunteer paraplegic patients produced a significant, prolonged depression of nociceptive reflexes without modifying monosynaptic reflexes. Moreover, this was verified clinically with a significant reduction in spasticity (Wurtman 1985).

Table 4. Intrathecal spinal drug delivery for control of spasticity

Authors, ref.	Etiology	N		Drug	Systems	
		Tested	Implanted		R	P
Erickson et al. (1985)	Post trauma	15	8	Morphine	–	8
	Demyelinating D	1	–			
Muller (1984)	Demyelinating D	3	–	Midazolan	–	–
Penn and Kroin (1984)	Post trauma	2	–	Baclofen	–	–
Lazorthes and Verdie (1985)	Post trauma	6	2	Morphine Baclofen	1	1
	Demyelinating D	5	2		2	–
	Cerebral palsy	2	2		1	1

– Rabbit studies with intrathecal lumbar administration of baclofen induced hind-foot flaccidity with small doses of 5 µg, whereas from 20 µg onwards, the front feet were also affected even though the animal remained alert (Kroin et al. 1984). This dose/effect relationship coincides with Yaksh studies.

– H. Muller successfully treated a spastic strychnine-induced cat through lumbar injections of a benzodiazepine, midazolan (0.15 mg/kg) together with intrathecal morphine administration.

A 2 mg dose of intrathecal morphine (Erickson et al. 1985) and a 10 mg–30 mg dose of epidural morphine (Struppler 1983) significantly improved both the pain and the spasticity of patients stricken with post-traumatic paraplegia and multiple sclerosis.

Currently clinical experience is limited to intrathecal administration of endogenous and exogenous ligands involved in the spinal motor reflex arc. These include morphine which is a mu-receptor agonist: baclofen which is a GABA B receptor agonist; and the benzodiazepines which are essentially GABA A agonists. Preliminary trials are summarized on Table 4. D. Erickson's experience is the most important, especially in the use of implantable pumps functioning at a continuous flow rate. This same author tested 16 patients, implanting 8 continuous-flow pumps (Infusaid) in 6 post-traumatic quadriplegic patients. A daily average dose of 2 mg controlled spasticity. There was a 1 to 17 month follow-up with no tolerance in those patients having iso-lated spasticity (Erickson et al. 1985).

– In 1984, Penn and Kroin reported their clinical experience on 2 paraplegic patients after injecting 25 µg–50 µg of baclofen. They suggest using implantable pumps for chronic intrathecal drug administration in these cases.

We have had experience with different spasticities of medullar and central origin (Lazorthes and Verdie 1985). Post-traumatic spasticities and spasticities secondary to demyelination and cerebral motor disability were tested. Six patients received implants: 4 with reservoirs and 2 with programmable D.A.S. pumps. Baclofen proved to be the most efficient drug for small dosage; however, the lack of a specific antagonist limited its applications. In fact, one of our patients received two serious baclofen overdoses causing loss of consciousness and hypoventilation which led us to discontinue its prescription. Small doses of morphine administered for chronic pain relief has proven extremely effective and much easier to control due to the fact that its effects are completely naloxone-reversible. Nevertheless, there are always the long-term risks with chronic opiate administration for patients having a long life expectancy.

Our interest in this method is based on its remarkable efficiency and its non-invasive, conservative, selective nature since it reduces spasticity without causing residual deficiencies in motor function. The future prospects for this method are undoubtedly very promising but we need further proof provided by rigorous, quantitative studies on spasticity, in particular double-blind and objective evaluations. Continuous drug administration requires extremely precise dosages that can be perfectly programmed; this is the advantage of programmable micropumps such as D.A.S. The inconvenience of this system, however, is that it is impossible to immediately stop perfusion in case of undesirable side effects. Nevertheless, the need for an implantable system permitting repeated bolus injections on demand is obvious, provided you can use weakly diluted solutions particularly with drugs as active as Baclofen.

Intracarotid Chemotherapy for Malignant Glioma

The cytotoxicity of antimitotics depends upon the duration of drug administration beyond a critical threshold. The brain and the cerebrospinal fluid are limited access sites and drugs delivered systemically do not reach the central nervous system in high concentrations. The carotid artery is the principal afferent for the majority of the supratentorial gliomas. It would then seem logical to use loco-regional antimitotic perfusion (Crafts et al. 1976). Fenstermacher and Cowles (1977) did monkey studies injecting equal amounts of BCNU marked with 14 carbon both intravenously and intracarotidly. Results showed that with the intracarotid approach the concentrations was 3 times greater in the homolateral brain than with the intravenous approach. In addition, the cerebral hemisphere that was perfused received 4 to 5 times the dosage that the contralateral hemisphere received.

The advantage of local chemotherapy treatment is that it not only avoids dispersing cytotoxic chemotherapy agents, but also avoids increasing their local plasmatic peak concentration. The indications are essentially cerebral malignant glioma (III and IV grade astrocytoma) with a unilateral hemispheric topography and secondary diffuse metastasis. Prior to chemotherapy surgical exeresis and conventional radiotherapy are performed. Chemotherapy is either given systematically, following radiotherapy, or only when there is a recurrence. Taking this concept and these indications as a basis initial results of intra-arterial chemotherapy using various chemical agents were reported (Greenberg 1984; Greenberg et al. 1981; Phillips et al. 1982). Some of these preliminary experiences are synthesized on Table 5. Most of the authors used the femoral approach with intra-arterial catheterisms during each chemotherapy session (normally every 2–3 months). The advantage of the percutaneous catheterism is to be as selective as possible (supra-ophthalmic), thus limiting the risks of neurotoxic retinal effects. Prior to chemotherapy, you may breakdown the blood-brain barrier by intravenous or even intra-arterial hypertonic mannitol perfusion followed by a rapid BCNU or cisplatinum perfusion (Neuwelt 1980).

Table 5. Regional chemotherapy for malignant brain tumors

Authors, ref.	Patients		Technic		Site	Drug
	GL	M	CATH	P		
Greenberg et al. (1981); Greenberg (1984)	30	–	30	–	I-A	BCNU
Dakhil et al. (1981)	7	–	–	7	I-V	Methotrexate
Philips et al. (1982)	6	–	–	6	I-A	BCNU, FUdR, Cisplatin
Feun et al. (1984)	20	10	30	–	I-A	Cisplatin
Stewart et al. (1984)	16	16	32		I-A	BCNU, Cisplatin
Beck et al. (1984)	–	3	–	3	I-Th	Methotrexate
Morantz (1984)	11	–	–	11	I-T	Bleomycin

GL = Gliomas grades III and IV; M = metastasis; CATH = femoral catheterism; P = pump; I-A = intraarterial; I-V = intraventricular; I-T = intratumoral; I-Th = intrathecal.

Few authors have used implantable pumps for this indication. In 1982 Phillips proposed a protocol which included a slow, continuous FudR perfusion (4.8 mg/day–6.5 mg/day) using an Infusaid 400 model pump for a 14–70 day period. A sideport permitted repeated bolus injections of either BCNU or of cisplatinum. No iatrogenic complications such as internal carotid thrombosis or transient ischemia were encountered during continuous external carotid perfusion. The risk of ocular toxicity is the major complication with this method (Greenberg 1984). It seems to be related to the alcoholic concentration of the BCNU solution; thus the importance of supra-ophthalmic catheterisms which reduce the risks of retinal toxicity. The second toxic risk is of a neurological sort, the risk being identical whether perfusion is done via the intra- or the supracarotid approach and regardless of which perfusion system is used. Nevertheless, implantable, continuous-flow perfusion pumps permit simultaneous perfusion of such "cell-cycle-specific" chemical agents as FudR or even radiosensitizers (BUdR). Another advantage of these slow continuous-flow pumps is to be able to maintain a high tissue concentration of a drug as necessary. Perhaps in the future we will be able to connect a slow-flow perfusion pump to an intra-arterial, supra ophthalmic catheter without encountering any iatrogenic complications.

Intraventricular chemotherapy has also been considered (Dakhil et al. 1981). Last of all, there is the direct intratumoral approach to chemotherapy whereby the catheter is implanted at the tumor site. There have been studies with rats as reported by Kroin and Penn (1982) administering cisplatin locally; however, it seems that tissue diffusion (maximum of 2 cm) limits such applications. The concept of local chemotherapy is tempting because it reduces the neurotoxic, neurological, ocular and more general risks to a minimum. Local micro-infusion may be the technique of the future. Clinical tests with man have already been performed using bleomycine in glioma cases (Morantz 1984).

Additional Applications

Progress made to our knowledge with neurochemical mechanisms concerning degenerative central nervous system diseases in the presence of a neuromediator deficit, has led to the development of new applications.

1. Alzheimer disease (A.D.). Recent data suggests there is a reduction in cholinergic cerebral activity in patients stricken with Alzheimer's disease (Harbaugh et al. 1984).

Concurrently, cerebral biopsies of A.D. patients showed a decrease in choline acetyl transpherase (ChAT), the specific cellular tracer of the cholinergic neurones, as well as a decrease in the acetylcholine synthesis (Coyle et al. 1983; Davies 1979).

Based on this cholinergic hypothesis of a decrease in muscarinic receptor activity in A.D. patients, Harbaugh et al. (1984) suggested a new therapy which consists of perfusing small doses of the muscarin agonist directly into the ventricular cerebrospinal fluid.

After analyzing toxic studies on dogs, we subjected 4 A.D. patients to a preliminary feasibility test followed by a cerebral biopsy. A constant-flow pump (INFUSAID) was connected to a catheter placed 3 times in the lateral ventricle and once in the great cistern. Bethanecolchloride was gradually perfused until the optimal therapeutic dose for each patient was obtained (between 0.05 mg/day and 0.7 mg/day). After an 8 month follow-up, we observed some spontaneous, reversible complications: initial nausea, and in one case, transient parkinsonian syndrome which retroceded after reducing the dose (0.6 mg/day to 0.4 mg/day). The therapeutic effects are encouraging, but we are still insufficiently documented. Subjective family reports noted improvement in the cognitive functions as well as in the patients' functional activity. Patients returned to their initial state after placebo perfusion. The efficiency of this treatment has yet to be confirmed by a longer follow-up, a quantitative evaluation of mental functions using standardized scales and a placebo controlled double-blind study. The prospects of this new therapeutic approach are essentially neuropharmacological; for example, in the case of Alzheimer's disease, our goal is to find a product capable of re-establishing the cholinergic activity where only one correction is needed (Yaksh and Rudy 1976).

2. The intrathecal administration of thyrotropine releasing hormone (TRH) in degenerative neurological diseases such as amyotrophic lateral sclerosis was proposed by T. L. Munsat (1984). It appears to be less toxic with a longer lasting effect than intravenous administration.

3. According to P. Ballantine (1984), the intraventricular administration of lithium has potential psychiatric applications. Kinetic animal studies showed that intraventricular drug delivery can be used to reduce extra-cerebral toxicity and the caudal neurotoxicity of lithium.

4. The intrathecal perfusion of glial GABA uptake inhibitors such as THPO was suggested in the treatment of obstinate epilepsy (Blackshear 1979).

5. Lastly, D. W. Roberts (1984) proposed using intracerebral drug administration at a precise action site through stereotaxy, not only for local chemotherapy as R. Penn did with cisplatinum, but also for the treatment of motor disorders affecting the basal nuclei.

Discussion

We will limit this discussion to implantable drug delivery systems and the standards of choice with respect to their applications for neurosurgery. As a general rule, regardless of the intended application, it is preferable to use implantable systems rather than portable external systems. No matter how small they may be, external pumps are poorly adapted to neurological indications due to their inherent restrictions on patients who are often disabled and highly susceptible to infection. The implantable system to be utilized must be:

1. Easy to implant and manipulate;
2. Composed of biocompatible material which does not modify the chemical stability of the drug being administered (Lazorthes et al. 1985).

Drugs contained in the pump reservoirs will remain there for various lengths of time, eventually beyond several months if necessary. During this period, the drug solution will be in contact with the different materials that constitute the pump. The materials mentioned on Table 6 have been selected or their biocompatibility qualities. Titanium is a noble metal widely used for cardiac and neurologic stimulator cases. It has been the material of choice for most of the implantable pump cases because it not only prevents corrosion, but also polymerization. Polysulfon resin is a plastic widely used for medical instruments because its specifications meet their stringent requirements. In fact, it has replaced all other plastics, in particular in the construction of implantable pumps due to its exceptional physical properties. Polysulfon resin is extremely inert and durable, notably resistant to autoclave (121°C) and ethylene oxide sterilization. It can be molded in complex shapes without additional assembly work which cuts costs considerably compared to the manufacture of metal products. Finally, most chemical agents have no effect on polysulfon. Medical-grade silicone elastomer is also highly appreciated for its biocompatibility and easy handling.

All these materials may cause the degradation of the drug solution administered regardless of what their physico-chemical qualities might be. For this reason drug stability tests must be performed on each implantable system. For our part, we have tested morphine stability in solutions destined for intrathecal administration using high-pressure liquid chromatography analysis (Lazorthes et al. 1985).

3. The system must be reliable not only from a programmation standpoint, but also from a security standpoint.
4. The pump should have a reservoir capacity sufficiently large to permit the patient significantly autonomy.

Table 6. Biocompatibles materials for implantable drug delivery systems

1. Metal	=	Titanium, stainless steel
2. Plastic	=	Polysulfone, teflon
3. Filters	=	Porus titanium
4. Wrapping	=	Silicon elastomer

5. An integral anti-bacterial filter must be included.
6. Its life expectancy must be sufficient, in particular if the energy source is also implanted.
7. The working mechanism should avoid all risks of involuntary administration.
8. Finally, the price should be acceptable, that is adapted to the patients life expectancy.

Other standards of choice are specific with respect to the particular indication, the administration site, and the drug to be delivered. Thus, concerning chemotherapy, the continuous flow pumps are perfectly suitable because they include a sideport which permits polychemotherapy in conjunction with continuous infusion of sequential bolus, in spite of their relatively imprecise flow rate. This type of pumps is much less suitable in cases requiring a precise delivery rate variable in time, because the flow rate cannot be modified. To compensate you must change the concentration of the drug to be delivered.

For antalogic purposes implantable mechanical pumps permitting the delivery of opiates "on demand" by repeated bulus injections are all the more appropriate given their low cost. The SECOR (Cordis) multidose reservoir which permits sequential bolus injections is an interesting solution. It would be the best solution if the reservoir was larger and if a sideport was incorporated, provided the cost remains low.

Implantable multiprogrammable electronic pumps provide the greatest precision and are perfectly suitable every time an active neuromediator drug is used entailing risks of side effects. Programmable pumps are appropriate for chronotherapy but, for the most part, their present-day use is limited due to their high cost; especially if we want to give the patient himself the means to stop perfusion in the event of undesirable side effects.

Be that as is may, the future prospects of local pharmacologic neurosurgery appear very promising. They will develop parallel with progress concerning the methods of action of endogenous and exogenous ligands in the neurophysiological mechanisms and their implications with degenerative cerebral disease. In our speciality, the progress of the future will be in the field of neurochemistry. From the biotechnological standpoint the development of implantable drug delivery systems must be adapted to the patients needs, strinving to improve their reliability and security while reducing their costs.

References

Ballantine P (1984) Intraventricular lithium infusion and potential applications in psychiatry. In: A Professional Briefing on "Totally implantable pumps", Isle of Palms, South Carolina, Sept 19–22

Beck DO (1984) Continuous infusion of methotrexate therapy of meningeal carcinomatosis. In: A Professional Briefing on "Totally implantable pumps", Isle of Palms, South Carolina, Sept 19–22

Behar M, Olshwang D, Magora F, Davidson JT (1979) Epidural morphine in treatment of pain. Lancet 1:527–528

Blackshear P (1979) Implantable drug delivery systems. Scientific American 241:66–73

Conseiller C, Menetrey D, Le Bars D, Besson JM (1972) Effect de la morphine sur les activités des interneurones de la couche V de Rexed de la corne dorsale chez le chat spinal. J Physiol (Paris) 65:220

Coyle JT, Price DL, Delong MR (1983) Alzheimer's disease: a disorder of cortical cholinergic innervation. Science 219:1184–1190

Crafts DC, Levin VA, Nielsen SA (1976) Intracarotid BCNU (NSC-409962): a toxicyty study in six rhesus monkeys. Cancer Treat Rep 60:541–545

Dakhil S, Ensminger W, Kindt G, Niedorhuber J, Chandler W, Greenburg H, Wheeler R (1981) Implanted system for intraventricular drug infusion in central nervous system tumors. Cancer Treat Rep 65:401–411

Davies P (1979) Neurotransmitter-related enzymes in senile dementia of the Alzheimer type. Brain Res 171:319–327

Duggan AW, Hall JG, Headley PM (1977) Suppression of transmission of nociceptive impulses by morphine: selective effects of morphine administered in the region of the substantia gelatinosa. Brit J Pharmacol 61:65–76

Erickson DL, Blacklock JB, Michaelson M, Sperling KB, Lo JN (1985) Control of spasticity by implantable continuous flow morphine pump. Neurosurgery 16:215–217

Fenstermacher JD, Cowles AL (1977) Theoretic limitation of intracarotid infusions in brain tumor chemotherapy. Cancer Treat Rep 61:519–526

Feun LG, Wallace S, Stewart DJ, Chuang WP, Yung WKA, Leavens ME, Burgess MA, Savaraj N, Benjamin RS, Young SE, Tang RA, Handel S, Mavligit G, Fields WS (1984) Intracarotid infusion of cis-diamminedichloroplatinum in the treatment of recurrent malignant brain tumors. Cancer 54:794–799

Greenberg HS (1984) Intra-arterial chemotherapy for malignant tumors of the central nervous system. In: A Professional Briefing on "Totally implantable pumps", Isle of Palms, South Carolina, Sept 19–22

Greenberg HS, Ensminger WD, Seeger JF, Kindt GW, Chandler F, Doan K, Dakhil SR (1981) Intra-arterial BCNU chemotherapy for the treatment of malignant gliomas of the central nervous system: a preliminary report. Cancer Treat Rep 65:803–810

Harbaugh RE, Roberts DW, Coombs DW, Saunders RL, Reeder TM (1984) Preliminary report: Intracranial cholinergic drug infusion in patients with Alzheimer's disease. Neurosurgery 15: 514–518

Kroin JS, Penn RD (1982) Intracerebral chemotherapy: chronic microinfusion of cisplatin. Neurosurgery 10:349–354

Kroin JS, Penn RD, Beissinger RL, Arzbaecher RC (1984) Reduced spinal reflexes following intrathecal baclofen in the rabbit. Exp Brain Res 54:191–194

Lazorthes Y, Verdié JC (1985) Neuropharmacologie en application intrathécale. In: Siegfried J, Lazorthes Y (eds) Neurochirurgie de l'infirmité motrice cérébrale. Neurochirurgie 31:95–101

Lazorthes Y, Gouardères Ch, Verdié JC, Montsarrat B, Bastide R, Campan L, Cros J (1980) Analgésie par injection intrathécale de morphine. Etude pharmacocinétique et application aux douleurs irréductibles. Neurochirurgie 26A:159–164

Lazorthes Y, Cros J, Verdié JC, Lagarrigue J, Bastide R, Boetto S (1982) Morphine intrarachidienne chronique dans le traitement des douleurs d'origine néoplasique. Médecine et Hygiène 40:1983–1989

Lazorthes Y, Verdié JC, Bastide R, Caute B, Clemente G (1985) Les systèmes implantables pour administration épidurale et intrathécale d'opioides. In: Besson JM, Lazorthes Y (eds) Analgésie spinale et opioides. Monographie INSERM

Leavens ME, Hill CS, Cech DA, Weyland JB, Weston JS (1982) Intrathecal and intraventricular morphine for pain in cancer patients: initial study. J Neurosurg 56:241–245

Lobato RD, Madrid JL, Fatela LV, Rivas JJ, Reig E, Lamas E (1983) Intraventricular morphine for control of pain in terminal cancer patients. J Neurosurg 59:627–633

Morantz MA (1984) Intraneoplastic chemotherapy in the treatment of primary brain tumors. In: A Professional Briefing on "Totally implantable pumps", Isle of Palms, South Carolina, Sept 19–22

Muller H (1984) Spinal benzodiazepines for treatment of spasticity. In: A Professional Briefing on "Totally implantable pumps", Isle of Palms, South Carolina, Sept 19–22

Munsat TL (1984) Intrathecal TRH in motor neuron diseases. In: A Professional Briefing on "Totally implantable pumps", Isle of Palms, South Carolina, Sept 19–22

Neuwelt EA, Frenkel EP, Rapoport S, Barnett P (1980) Effect of osmotic blood-brain barrier disruption on methotrexate pharmacokinetics in the dog. Neurosurgery 7:36–43

Nurchi G (1984) Use of intraventricular and intrathecal morphine in intractable pain associated with cancer. Neurosurgery 15:801–803

Onofrio BM, Yaksh TL, Arnold PG (1981) Continuous low-dose intrathecal morphine administration in the treatment of chronic pain of malignant origin. Mayo Clin Proc 56:516–520

Penn RD, Kroin JS (1984) Intrathecal baclofen alleviates spinal cord spasticity. Lancet, p 1078

Penn RD, Paice JA, Gottschalk W, Ivankovich AD (1984) Cancer pain relief using chronic morphine infusion: early experience with a programmable implanted drug pump. J Neurosurg 61: 302–306

Pert CB, Snyder SH (1973) Opiate receptor: demonstration in nervous tissue. Science 179:1011–1014

Phillips TW, Chandler WF, Kindt GW, Ensminger WD, Greenberg HS, Seeger JF, Doan KM, Gyves JW (1982) New implantable continuous administration and bolus dose intracarotid drug delivery system for the treatment of malignant gliomas. Neurosurgery 11:213–218

Poletti CE, Cohen AM, Todd DP (1981) Cancer pain relieved by long-term epidural morphine with permanent indwelling systems for self-administration. Neurosurgery 55:581–584

Poletti CE, Sweet WH, Schmidek HH, Pilon RN (1983) Pain control with implantable systems for the long-term infusion of intraspinal opioids in man. In: Schmidek HH, Sweet WH (eds) Operative neurosurgical techniques. Grune and Stratton Publishers, 77:1199–1210

Roberts DW (1984) Stereotactic drug administration to the central nervous system. In: A Professionnal Briefing on "Totally implantable pumps", Isle of Palms, South Carolina, Sept 19–22

Roquefeuil B, Benezech J, Blanchet P, Batier C, Frerebeau P, Gros C (1984) Intraventricular administration of morphine in patients with neoplastic intractable pain. Surg Neurol 21:155–158

Saunders RL, Coombs DW (1983) Dartmouth-Hitchock Medical Center experience with continuous intraspinal narcotic analgesia. In: Schmidek HH, Sweet WH (eds) Operative neurosurgical techniques. Grune and Stratton, 77:1211–1212

Schousboe A (1984) Glial GABA uptake inhibitors and epilepsy. In: A Professional Briefing on "Totally implantable pumps", Isle of Palms, South Carolina, Sept 19–22

Stewart DJ, Grahovac Z, Benoit B, Addison D, Richard MT, Dennery J, Hugenholtz H, Russell N, Peterson E, Maroun JA, Vandenberg T, Hopkins HS (1984) Intracarotid chemotherapy with a combination of 1,3-bis(2-chloroethyl)-1-nitrosourea (BCNU), cis-diaminedichloroplatinum (cisplatin), and 4′-0-(4,6-0-2-thenylidene-beta-d-glucopyranosyl) epipodophyllotoxin (VM-26) in the treatment of primary and metastatic brain tumors. Neurosurgery 15:828–833

Struppler A (1983) Epidural morphine in spasticity. Case report. In: Congress of "Endorphines, Neurohormones and Transmitters", Garmisch-Partenkirchen, June 20

Thiebaut JB, Blond S, Farcot JM, Thurel C, Matge G, Schach G, Meynadier J, Bucheit F (1985) La morphine par voie intraventriculaire dans le traitement des douleurs néoplasiques. Mèdicine et Hygiène 43:636–646

Wang JK (1977) Analgesic effect of intrathecally administered morphine. Reg Anesth 4:2–3

Wang JK (1985) Intrathecal morphine for intractable pain secondary to cancer of pelvic organs. Pain 21:99–102

Willer JL, Bussel B (1980) Evidence for a direct spinal mechanism in morphine induced inhibition of nociceptive reflexes in humans. Brain Res 187:212–215

Wurtman R (1985) La maladie d'Alzheimer. Pour la Science 46–55

Yaksh TL, Rudy TA (1976) Analgesia mediated by a direct spinal action of narcotics. Science 192: 1357–1358

Acknowledgements

This publication is partially supported by C.N.R., Associazione "Paolo Zorzi".

We thanks also Mrs. Renata Gianferrari for the editing secretarial help.

The Symposium has been made possible also with financial support of the following Companies: Aesculap; Artemide; Battaglia Rangoni; Bracco; Codman; Comesa; FK Medica; M.G. Lorenzatto-Meditek; Medtronic; Rank Precision; Schiapparelli Medtronic; Tekmed; B. Tonon; Valfivre; Zeiss, Associazione A.R.I.N., and the Bank Cassa di Risparmio delle Province Lombarde (CARIPLO).

Subject Index

Advances in Neurosurgery

Volume 15: **R. Wüllenweber, M. Klinger, M. Brock** (Eds.)

Regulation of Cerebral Blood Flow and Metabolism. Neurosurgical Treatment of Epilepsy. Rehabilitation in Neurosurgery

1987. 102 figures. XII, 329 pages. ISBN 3-540-17402-8

Volume 14: **H. Wenker, M. Klinger, M. Brock, F. Reuter** (Eds.)

Spinal Cord Tumors. Experimental Neurosurgery. Neurosurgical Intensive Care

1986. 199 figures, 65 tables. XII, 386 pages. ISBN 3-540-16360-3

Volume 13: **H. Dietz, M. Brock, M. Klinger** (Eds.)

Extra-Intracranial Vascular Anastomoses, Microsurgery at the Edge of the Tentorium

1985. 179 figures, 71 tables. XII, 351 pages. ISBN 3-540-15615-1

Volume 12: **W. Piotrowski, M. Brock, M. Klinger** (Eds.)

CNS Metastases, Neurosurgery in the Aged

1984. 129 figures, 130 tables. XIV, 346 pages.
ISBN 3-540-12832-8

Volume 11: **H.-P. Jensen, M. Brock, M. Klinger** (Eds.)

Acute Non-Traumatic Intracranial Bleedings

Posterior Fossa Tumors in Infancy

1983. 186 figures, 122 tables. XII, 405 pages. ISBN 3-540-12538-8
Distribution rights for Japan: Nankodo Co., Ltd., Tokyo

Springer-Verlag
Berlin Heidelberg New York
London Paris Tokyo

Springer

J. D. Miller, Edinburgh; **G. M. Teasdale, J. O. Rowan, S. L. Galbraith, A. D. Mendelow**, Glasgow (Eds.)

Intracranial Pressure VI

1986. 360 figures, 127 tables. XXIV, 798 pages.
ISBN 3-540-16197-X

M. Samii (Ed.)

Surgery in and around the Brain Stem and the Third Ventricle. Anatomy. Pathology. Neurophysiology. Diagnosis. Treatment

1986. 370 figures. XXIII, 599 pages. ISBN 3-540-16581-9

M. Samii, J. Brihaye (Eds.)

Traumatology of the Skull Base. Anatomy, Clinical, and Radiological Diagnosis, Operative Treatment

1983. 103 figures. XIV, 240 pages. (First International Congress of the Skull Base Study Group).
ISBN 3-540-12528-0

Contents: Pathophysiology of Skull Base Injuries. – Evaluation and General Management of Skull Base Injuries. – Olfactory Nerve and Tract. – Optic Nerve and Orbit. – Facial Nerve. – Vestibulocochlear Nerve and Hearing Mechanisms. – Vascular Injuries. – Craniocervical Junction. – Subject Index.

J. Schramm, S. Jones (Eds.)

Spinal Cord Monitoring

1985. 139 figures, 63 tables. XII, 329 pages.
ISBN 3-540-15774-3

Springer-Verlag
Berlin Heidelberg New York
London Paris Tokyo

Springer

SPRINGER NATURE

GPSR Compliance

The European Union's (EU) General Product Safety Regulation (GPSR) is a set of rules that requires consumer products to be safe and our obligations to ensure this.

If you have any concerns about our products, you can contact us on ProductSafety@springernature.com

In case Publisher is established outside the EU, the EU authorized representative is:

Springer Nature Customer Service Center GmbH
Europaplatz 3
69115 Heidelberg, Germany

FSC
www.fsc.org
MIX
Papier | Fördert gute Waldnutzung
FSC® C083411

Kolibri 360 GmbH
Ferdinand-Jühlke-Straße 7
99095 Erfurt, Deutschland
produktsicherheit@kolibri360.de